NANOTECHNOLOGY SCIENCE AND TECHNOLOGY

NANOMANUFACTURING

AN EMERGING MEGATREND AND IMPLICATIONS FOR THE UNITED STATES

NANOTECHNOLOGY SCIENCE AND TECHNOLOGY

Additional books in this series can be found on Nova's website under the Series tab.

Additional e-books in this series can be found on Nova's website under the e-book tab.

NANOTECHNOLOGY SCIENCE AND TECHNOLOGY

NANOMANUFACTURING

AN EMERGING MEGATREND AND IMPLICATIONS FOR THE UNITED STATES

BRAD D. MCREYNOLDS
EDITOR

nova publishers
New York

Library of Congress Cataloging-in-Publication Data

ISBN: 978-1-63117-639-5

Published by Nova Science Publishers, Inc. † New York

Contents

PREFACE

Nanoscale science, engineering, and technology—commonly referred to collectively as nanotechnology—is believed by many to offer extraordinary economic and societal benefits. Congress has demonstrated continuing support for nanotechnology and has directed its attention primarily to three topics that may affect the realization of this hoped for potential federal research and development (R&D) in nanotechnology; U.S. competitiveness; and environmental, health, and safety (EHS) concerns. This book provides an overview of these topics, and two others which are nanomanufacturing, and public understanding of and attitudes toward nanotechnology.

Chapter 1 - Nanoscale science, engineering, and technology—commonly referred to collectively as nanotechnology—is believed by many to offer extraordinary economic and societal benefits. Congress has demonstrated continuing support for nanotechnology and has directed its attention primarily to three topics that may affect the realization of this hoped for potential: federal research and development (R&D) in nanotechnology; U.S. competitiveness; and environmental, health, and safety (EHS) concerns. This report provides an overview of these topics and two others: nanomanufacturing and public understanding of and attitudes toward nanotechnology.

The development of this emerging field has been fostered by significant and sustained public investments in nanotechnology R&D. Nanotechnology R&D is directed toward the understanding and control of matter at dimensions of roughly 1 to 100 nanometers. At this size, the properties of matter can differ in fundamental and potentially useful ways from the properties of individual atoms and molecules and of bulk matter. Since the launch of the National Nanotechnology Initiative (NNI) in 2000 through FY2013, Congress has appropriated approximately $18 billion for nanotechnology R&D. President Obama has requested $1.7 billion in NNI funding for FY2014. More than 60 nations have established similar programs. In 2010, total annual global public R&D investments reached an estimated $8.2 billion, complemented by an estimated private sector investment of $9.6 billion. Data on economic outputs used to assess competitiveness in mature technologies and industries, such as revenues and market share, are not available for assessing nanotechnology. Alternatively, data on inputs (e.g., R&D expenditures) and non-financial outputs (e.g., scientific papers, patents) may provide insight into the current U.S. position and serve as bellwethers of future competitiveness. By these criteria, the United States appears to be the overall global leader in nanotechnology, though some believe the U.S. lead may not be as large as it was for previous emerging technologies.

Some research has raised concerns about the safety of nanoscale materials. There is general agreement that more information on EHS implications is needed to protect the public and the environment; to assess and manage risks; and to create a regulatory environment that fosters prudent investment in nanotechnology-related innovation. Nanomanufacturing—the bridge between nanoscience and nanotechnology products—may require the development of new technologies, tools, instruments, measurement science, and standards to enable safe, effective, and affordable commercial-scale production of nanotechnology products. Public understanding and attitudes may also affect the environment for R&D, regulation, and market acceptance of products incorporating nanotechnology.

In 2003, Congress enacted the 21st Century Nanotechnology Research and Development Act (P.L. 108-153) providing a legislative foundation for some of the activities of the NNI, addressing concerns, establishing programs, assigning agency responsibilities, and setting authorization levels. Efforts to reauthorize the act have been unsuccessful. As of the date of this report, no reauthorization legislation had been introduced in the 113th Congress. In October 2013, the ranking member of the House Committee on Science, Space, and Technology circulated a draft reauthorization of the America COMPETES Act that included a "Reauthorization of the National Nanotechnology Initiative" subtitle. The majority version of the 2013 America COMPETES Act reauthorization bill does not include a nanotechnology reauthorization provision.

Chapter 2 - Nanotechnology has been defined as the control or restructuring of matter at the atomic and molecular levels in the size range of about 1–100 nanometers (nm); 100 nm is about 1/1000th the width of a hair.

The U.S. National Nanotechnology Initiative (NNI), begun in 2001 and focusing primarily on R&D, represents a cumulative investment of almost $20 billion, including the request for fiscal year 2014. As research continues and other nations increasingly invest in R&D, nanotechnology is moving from the laboratory to commercial markets, mass manufacturing, and the global marketplace—a trend with potential future import that some compare to history's introduction of technologies with major economic and societal impact, such as plastics and even electricity. Today, burgeoning markets, innovation systems, and nanomanufacturing activities are increasingly competitive in a global context—and the potential EHS effects of nanomanufacturing remain largely unknown.

At the July 2013 forum, participants from industry, government, and academia discussed the future of nanomanufacturing; investments in nanotechnology R&D and challenges to U.S. competitiveness; ways to enhance U.S. competitiveness; and EHS concerns. Participants reviewed a summary of forum discussions, and two experts (who did not attend the forum) independently reviewed a draft of this report. Their comments were incorporated in this report as appropriate.

In: Nanomanufacturing
Editor: Brad D. McReynolds
ISBN: 978-1-63117-639-5

Chapter 1

NANOTECHNOLOGY: A POLICY PRIMER*

John F. Sargent Jr.

SUMMARY

Nanoscale science, engineering, and technology—commonly referred to collectively as nanotechnology—is believed by many to offer extraordinary economic and societal benefits. Congress has demonstrated continuing support for nanotechnology and has directed its attention primarily to three topics that may affect the realization of this hoped for potential: federal research and development (R&D) in nanotechnology; U.S. competitiveness; and environmental, health, and safety (EHS) concerns. This report provides an overview of these topics and two others: nanomanufacturing and public understanding of and attitudes toward nanotechnology.

The development of this emerging field has been fostered by significant and sustained public investments in nanotechnology R&D. Nanotechnology R&D is directed toward the understanding and control of matter at dimensions of roughly 1 to 100 nanometers. At this size, the properties of matter can differ in fundamental and potentially useful ways from the properties of individual atoms and molecules and of bulk matter. Since the launch of the National Nanotechnology Initiative (NNI) in 2000 through FY2013, Congress has appropriated approximately $18 billion for nanotechnology R&D. President Obama has requested $1.7 billion in

* This is an edited, reformatted and augmented version of a Congressional Research Service publication, CRS Report for Congress RL34511, from www.crs.gov, prepared for Members and Committees of Congress, dated December 16, 2013.

NNI funding for FY2014. More than 60 nations have established similar programs. In 2010, total annual global public R&D investments reached an estimated $8.2 billion, complemented by an estimated private sector investment of $9.6 billion. Data on economic outputs used to assess competitiveness in mature technologies and industries, such as revenues and market share, are not available for assessing nanotechnology. Alternatively, data on inputs (e.g., R&D expenditures) and non-financial outputs (e.g., scientific papers, patents) may provide insight into the current U.S. position and serve as bellwethers of future competitiveness. By these criteria, the United States appears to be the overall global leader in nanotechnology, though some believe the U.S. lead may not be as large as it was for previous emerging technologies.

Some research has raised concerns about the safety of nanoscale materials. There is general agreement that more information on EHS implications is needed to protect the public and the environment; to assess and manage risks; and to create a regulatory environment that fosters prudent investment in nanotechnology-related innovation. Nanomanufacturing—the bridge between nanoscience and nanotechnology products—may require the development of new technologies, tools, instruments, measurement science, and standards to enable safe, effective, and affordable commercial-scale production of nanotechnology products. Public understanding and attitudes may also affect the environment for R&D, regulation, and market acceptance of products incorporating nanotechnology.

In 2003, Congress enacted the 21st Century Nanotechnology Research and Development Act (P.L. 108-153) providing a legislative foundation for some of the activities of the NNI, addressing concerns, establishing programs, assigning agency responsibilities, and setting authorization levels. Efforts to reauthorize the act have been unsuccessful. As of the date of this report, no reauthorization legislation had been introduced in the 113th Congress. In October 2013, the ranking member of the House Committee on Science, Space, and Technology circulated a draft reauthorization of the America COMPETES Act that included a "Reauthorization of the National Nanotechnology Initiative" subtitle. The majority version of the 2013 America COMPETES Act reauthorization bill does not include a nanotechnology reauthorization provision.

Overview

Congress continues to demonstrate interest in and support for nanotechnology due to what many believe is its extraordinary potential for delivering economic growth, high-wage jobs, and other societal benefits to the nation. To date, the Science and Technology Committee in the House and Senate Committee on Commerce, Science, and Transportation have directed their attention primarily to three topics that may affect the United States' realization of this hoped for potential: federal research and development (R&D) investments under the National Nanotechnology Initiative (NNI); U.S. international competitiveness; and environmental, health, and safety (EHS) concerns. This report provides a brief overview of these topics—which are discussed in greater detail in other CRS reports[1]—and two other subjects of interest to Congress: nanomanufacturing and public attitudes toward, and understanding of, nanotechnology.

Nanotechnology research and development is directed toward the understanding and control of matter at dimensions of roughly 1 to 100 nanometers. At this size, the physical, chemical, and biological properties of materials can differ in fundamental and potentially useful ways from the properties of individual atoms and molecules, on the one hand, or bulk matter, on the other hand.

In 2000, President Clinton launched the NNI to coordinate federal R&D efforts and promote U.S. competitiveness in nanotechnology. Congress first funded the NNI in FY2001 and has provided increased appropriations for nanotechnology R&D in each subsequent year. In 2003, Congress enacted the 21st Century Nanotechnology Research and Development Act (P.L. 108-153). The act provided a statutory foundation for the NNI, established programs, assigned agency responsibilities, authorized agency funding levels for FY2005 through FY2008, and initiated research to address key issues.

Federal R&D investments are focused on advancing understanding of fundamental nanoscale phenomena and on developing nanomaterials, nanoscale devices and systems, instrumentation, standards, measurement science, and the tools and processes needed for nanomanufacturing. NNI appropriations also fund the construction and operation of major research facilities and the acquisition of instrumentation. Finally, the NNI supports research directed at identifying and managing potential environmental, health, and safety impacts of nanotechnology, as well as its ethical, legal, and societal implications.

Most current applications of nanotechnology are evolutionary in nature, offering incremental improvements in existing products and generally modest economic and societal benefits. For example, nanotechnology is being used in automobile bumpers, cargo beds, and step-assists to reduce weight, increase resistance to dents and scratches, and eliminate rust; in clothes to increase stain- and wrinkle-resistance; and in sporting goods, such as baseball bats and golf clubs, to improve performance.

In the longer term, nanotechnology may deliver revolutionary advances with profound economic and societal implications. Potential applications discussed by the technology's proponents involve various degrees of speculation and varying time-frames. The examples below suggest areas where such possible revolutionary advances may emerge, and early research and development efforts that may provide insights into how such advances may be achieved.

- *Detection and treatment technologies for cancer and other deadly diseases.* Current nanotechnology disease detection efforts include the development of sensors that can identify biomarkers, such as altered genes, that may provide an early indicator of cancer. One approach uses carbon nanotubes and nanowires to identify the unique molecular signals of cancer biomarkers. Another approach uses nanoscale cantilevers—resembling a row of diving boards—treated with molecules that bind only with cancer biomarkers. When these molecules bind, the additional weight bends the cantilevers indicating the presence and concentration of these biomarkers. Nanotechnology holds promise for showing the presence, location, and/or contours of cancer, cardiovascular disease, or neurological disease. Current R&D efforts employ metallic, magnetic, and polymeric nanoparticles with strong imaging characteristics attached to an antibody or other agent that binds selectively with targeted cells. The imaging results can be used to guide surgical procedures and to monitor the effectiveness of non-surgical therapies in killing the disease or slowing its growth. Nanotechnology may also offer new cancer treatment approaches. For example, nanoshells with a core of silica and an outer metallic shell can be engineered to concentrate at cancer lesion sites. Once at the sites, a harmless energy source (such as near-infrared light) can be used to cause the nanoshells to heat, killing the cancer cells they are attached to.[2] Another treatment approach targets delivery of tiny amounts of a chemotherapy drug to cancer cells. In this approach the drug is encapsulated inside a nanoshell that is engineered to bind with an antigen on the cancer cell. Once bound, the nanoshell dissolves, releasing the chemotherapy drug, killing the cancer cell. Such a targeted delivery approach could reduce the amount of chemotherapy drug needed to kill the cancer cells, reducing the side effects of chemotherapy.[3]

- *Clean, inexpensive, renewable power through energy creation, storage, and transmission technologies.* Nanoscale semiconductor catalysts and additives show promise for improving the production of hydrogen from water using sunlight. The optical properties of these nanoscale catalysts allow the process to use a wider spectrum of sunlight. Similarly, nanostructured photovoltaic devices (e.g., solar panels) may improve the efficiency of converting sunlight into electricity by using a wider spectrum of sunlight. Improved hydrogen storage, a key challenge in fuel cell applications, may be achieved by tapping the chemical properties and large surface area of certain nanostructured materials. In addition, carbon nanotube fibers have the potential for reducing energy transmission losses from approximately 7% (using copper wires) to 6% (using carbon nanotube fibers), an equivalent annual energy savings in the United States of 24 million barrels of oil.[4]
- *Universal access to clean water.* Nanotechnology water desalination and filtration systems may offer affordable, scalable, and portable water filtration systems. Filters employing nanoscale pores work by allowing water molecules to pass through, but prevent larger molecules, such as salt ions and other impurities (e.g., bacteria, viruses, heavy metals, and organic material), from doing so. Some nanoscale filtration systems also employ a matrix of polymers and nanoparticles that serve to attract water molecules to the filter and to repel contaminants.[5]
- *High-density memory devices.* A variety of nanotechnology applications may hold the potential for improving the density of memory storage and accelerate access speed to stored data.[6]
- *Higher crop yield and improved nutrition.* Higher crop yield might be achieved using nanoscale sensors that detect the presence of a virus or disease-infecting particle. Early, location-specific detection may allow for rapid and targeted treatment of affected areas, increasing yield by preventing losses.[7] Nanotechnology also offers the potential for improved nutrition. Some companies are exploring the development of nanocapsules that release nutrients targeted at specific parts of the body at specific times.[8]

- *Self-healing materials.* Nanotechnology may offer approaches that enable materials to "self-heal" by incorporating, for example, nanocontainers of a repair substance (e.g., an epoxy) throughout the material. When a crack or corrosion reaches a nanocontainer, the nanocontainer could be designed to open and release its repair material to fill the gap and seal the crack.[9]
- *Sensors that can warn of minute levels of toxins and pathogens in air, soil, or water.* Microfluidic and nanocantilever sensors (discussed earlier) may be engineered to detect specific pathogens (e.g., bacteria, virus) or toxins (e.g., sarin gas, hydrogen cyanide) by detecting their unique molecular signals or through selective binding with an engineered nanoparticle.[10]
- *Environmental remediation of contaminated sites.* The high surface-tovolume ratio, high reactivity, and small size of some nanoscale particles (e.g., nanoscale iron) may offer more effective and less costly solutions to environmental contamination. By injecting engineered nanoparticles into the ground, these characteristics can be employed to enable the particles to move more easily through a contaminated site and bond more readily with targeted contaminants.[11]

Nanotechnology is also expected by some to make substantial contributions to federal missions such as national defense, homeland security, and space exploration and commercialization.

U.S. private sector nanotechnology R&D is now estimated to be twice that of public funding. In general, the private sector's efforts are focused on translating fundamental knowledge and prototypes into commercial products; developing new applications incorporating nanoscale materials; and developing technologies, methods, and systems for commercial-scale manufacturing.

Many other nations and firms around the world are also making substantial investments in nanotechnology to reap its potential benefits. Between 2001 and 2004, more than 60 countries established nanotechnology programs at the national level.[12]

With so much potentially at stake, some Members of Congress have expressed interest and concerns about the U.S. competitive position in nanotechnology R&D and success in translating R&D results to commercial products. This has led to an increased focus on potential barriers to commercialization efforts, including the readiness of technologies, systems, and processes for large-scale nanotechnology manufacturing; potential environmental, health, and safety (EHS) effects of nanoscale materials; public understanding and attitudes toward nanotechnology; and other related issues.

This report provides an overview of the NNI, macro-level view of federal R&D investments in nanotechnology, U.S. competitiveness in nanotechnology, and EHS-related issues.

THE NATIONAL NANOTECHNOLOGY INITIATIVE

President Clinton launched the National Nanotechnology Initiative in 2000, establishing a multi-agency program to coordinate and expand federal efforts to advance the state of nanoscale science, engineering, and technology, and to position the United States to lead the world in its development and commercialization. The NNI is comprised of 15 federal agencies that receive appropriations to conduct and fund nanotechnology R&D and 12 other federal agencies with responsibilities for health, safety, and environmental regulation; trade; education; training; intellectual property; international relations; and other areas that might affect nanotechnology. The Environmental Protection Agency and the Food and Drug Administration both conduct nanotechnology R&D and have regulatory responsibilities.

Congress has played a central role in the NNI, providing appropriations for the conduct of nanotechnology R&D (discussed below), establishing programs, and creating a legislative foundation for some of the activities of the NNI through enactment of the 21st Century Nanotechnology Research and Development Act of 2003. The act authorized appropriations for FY2005 through FY2008 for five NNI agencies–the National Science Foundation (NSF), Department of Energy (DOE), National Aeronautics and Space Administration (NASA), Department of Commerce (DOC) National Institute of Standards and Technology (NIST), and Environmental Protection Agency (EPA). Congress remains actively engaged in the NNI.

While many provisions of the 21st Century Nanotechnology Research and Development Act have no sunset provision, FY2008 was the last year of agency authorizations included in the act. Legislation to amend and reauthorize the act was introduced in the House (H.R. 5940, 110th Congress) and the Senate (S. 3274, 110th Congress) in the 110th Congress. The House passed H.R. 5940 by a vote of 407-6; the Senate did not act on S. 3274. In January 2009, H.R. 554 (111th Congress), the National Nanotechnology Initiative Amendments Act of 2009, was introduced in the 111th Congress. The act contained essentially the same provisions as H.R. 5940. In February 2009, the House passed the bill by voice vote under a suspension of the rules. The bill was referred to the Senate Committee on Commerce, Science, and Transportation; no further action was taken. On May 7, 2010, the House Committee on Science and Technology reported the America COMPETES Reauthorization Act of 2010 (H.R. 5116, 111th Congress) which included, as Title I, Subtitle A, of the National Nanotechnology Initiative Amendments Act of 2010. This title was removed prior to enactment.[13] No reauthorization bill was introduced in the 112th Congress.

In October 2013, the ranking member of the House Committee on Science, Space, and Technology circulated a draft reauthorization of the America COMPETES Act that included a "Reauthorization of the National Nanotechnology Initiative" subtitle.[14] The majority version of the America COMPETES Act reauthorization does not include a nanotechnology reauthorization provision.

Structure

The NNI is coordinated within the White House through the National Science and Technology Council's NSET subcommittee. The NSET subcommittee is comprised of representatives from 27 federal agencies, White House Office of Science and Technology Policy (OSTP), and Office of Management and Budget.[15] The NSET subcommittee has established four working groups: the National Environmental and Health Implications (NEHI), National Innovation and Liaison with Industry (NILI), Global Issues in Nanotechnology (GIN), Nanomanufacturing, and Nanotechnology Public Engagement and Communications (NPEC) working groups. The National Nanotechnology Coordination Office (NNCO) provides administrative and technical support to the NSET subcommittee.

Funding

Funding for the NNI is provided through appropriations to each of the NNI-participating agencies. The NNI has no centralized funding. Overall NNI funding is calculated by aggregating the nanotechnology-related expenditures of each NNI agency. Funding remains concentrated in the original six NNI agencies,[16] which account for approximately 96% of NNI funding in FY2012.

The NNI funds fundamental and applied nanotechnology R&D, multidisciplinary centers of excellence, and key research infrastructure. It also supports efforts to address societal implications of nanotechnology, including ethical, legal, EHS, and workforce issues.

For FY2012, Congress appropriated an estimated $1.857 billion for nanotechnology R&D, four times the $464 million appropriated for nanotechnology R&D in 2001. The NNCO was not able to specify FY2013 actual NNI funding in its FY2014 budget supplement due to the late resolution of the federal budget process. FY2013 funding remained unavailable at the time of this report. In total, Congress has appropriated approximately $18 billion for the NNI from FY2001 to FY2013. President Obama has requested $1.702 billion for nanotechnology R&D in FY2014, a $155 million (8.4%) decrease below the actual FY2012 funding level of $1.857 billion. The chronology of NNI funding is detailed in *Table 1*.

SELECTED ISSUES

U.S. Competitiveness

Nanotechnology is largely still in its infancy. Accordingly, measures such as revenues, market share, and global trade statistics—which are often used to assess and track U.S. competitiveness in other more mature technologies and industries—are not available for assessing the relative U.S. position internationally in nanotechnology. To date, the federal government does not collect data on nanotechnology-related revenues, trade, or employment, nor are comparable international government data available. Nevertheless, many experts believe that the United States is the global leader in nanotechnology. However, some of these experts believe that in contrast to many previous emerging technologies—such as semiconductors, satellites, software, and biotechnology—the U.S. lead is narrower, and the investment level, scientific and industrial infrastructure, technical capabilities, and science and engineering workforces of other nations are more substantial than in the past.

In the absence of comprehensive and reliable economic output data (e.g., revenues, market share, trade), indicators such as inputs (e.g., public and private research investments) and non-financial outputs (e.g., scientific papers, patents) have been used to gauge a nation's competitive position in emerging technologies. By these measures (discussed below), the United States appears to lead the world, generally, in nanotechnology.

Table 1. NNI Funding, by Agency
(in millions of current dollars)

Agency	FY 2001 Actual	FY 2002 Actual	FY 2003 Actual	FY 2004 Actual	FY 2005 Actual	FY 2006 Actual	FY 2007 Actual	FY 2008 Actual	FY 2009 Actual	FY 2009 ARRA	FY 2010 Actual	FY 2011 Actual	FY 2012 Actual	FY 2014 Request
National Institutes of Health (DHHS)[a]	40	59	78	106	165	192	215	305	343	73	457	409	456	461
National Science Foundation	150	204	221	256	335	360	389	409	409	101	429	485	466	431
Department of Energy[b]	88	89	134	202	208	231	236	245	333	293	374	346	314	370
Department of Defense[c]	125	224	220	291	352	424	450	460	459		440	425	426	217
National Institute of Standards and Technology (DOC)	33	77	64	77	79	78	88	86	93	43	115	96	95	102
National Aeronautics and Space Administration	22	35	36	47	45	50	20	17	14		20	17	19	18
Environmental Protection Agency	5	6	5	5	7	5	8	12	12		18	17	18	17
Other Agencies	1	3	2	5	9	13	19	22	40		62	32	64	87
TOTAL[d]	464	697	760	989	1,200	1,351	1,425	1,554	1,702	511	1913	1,845	1,857	1,767

Source: NNI website. http://www.nano.gov/. Figures for FY2012 and FY2014 from The National Nanotechnology Initiative: Supplement to the President's FY2014 Budget, National Science and Technology Council, Executive Office of the President, May 2013.

[a] According to NIH, the agency has adopted the Research, Condition, and Disease Categorization (RCDC) system to provide more consistent and transparent information to the public about NIH research. The shift to the RCDC process of categorization changes the way individual research projects are assigned to categories. This change will result in some differences in total dollar amounts between the 2008 reports and those issued in previous years. Any difference, whether an increase or decrease in funding levels, does not necessarily reflect a change in the amount of money the NIH received from Congress or a change in the actual content of the NIH research portfolio. For more information, please go to: http://report.nih.gov/rcdc/reasons/default.aspx.

[b] According to NSTC, funding levels for DOE include the combined budgets of the Office of Science, the Office of Energy Efficiency and Renewable Energy, the Office of Fossil Energy, and the Advanced Research Projects Agency for Energy.

[c] According to NSTC, the Department of Defense budgets for FY2006-FY2011 include congressionally directed funding outside the NNI plan. The extent to which such funding is included or not included in reporting of funding in earlier fiscal years is uncertain.

[d] Numbers may not add due to rounding of agency budget figures.

However, R&D investments, scientific papers, and patents may not provide reliable indicators of the United States' current or future competitive position. Scientific and technological leadership may not necessarily result in commercial leadership or national competitiveness for a variety of reasons:

- Basic research in nanotechnology may not translate into viable commercial applications.
- Basic research is generally available to all competitors.
- U.S.-based companies may conduct production and other work outside of the United States.
- U.S.-educated foreign students may return home to conduct research and create new businesses.
- U.S. companies with leading-edge nanotechnology capabilities and/or intellectual property may be acquired by foreign competitors.
- U.S. policies or other factors may prohibit nanotechnology commercialization, make it unaffordable, or make it less attractive than foreign alternatives.
- Aggregate national data may be misleading as countries may establish global leadership in niche areas of nantoechnology.

With these caveats, the following section reviews input and non-economic output measures as indicators of the U.S. competitive position in nanotechnology.

Global Funding

The United States has led, and continues to lead, all nations in known public investments in nanotechnology R&D, though the estimated U.S. share of global public investments has fallen as other nations have established similar programs and increased funding. In 2011, Lux Research, an emerging technologies consulting firm, estimated total (public and private) global nanotechnology funding for 2010 to be approximately $17.8 billion with corporate R&D accounting for a majority of funding for the first time.[17] Cientifica, a privately held nanotechnology business analysis and consulting firm, estimated global public investments in nanotechnology in 2010 to be approximately $10 billion per year, with cumulative global public investments through 2011 reaching approximately $67.5 billion. Cientifica also concluded that the United States had fallen behind both Russia and China in nanotechnology R&D funding on a purchasing power parity (PPP) basis (which takes into account the price of goods and services in each nation), but still leads the world in real dollar terms (adjusted on a currency exchange rate basis).[18]

Private investments in nanotechnology R&D come from two primary sources, corporations and venture capital investors. Lux Research estimated that total global private sector nanotechnology funding had risen from $9.2 billion in 2009 to $9.6 billion in 2010, while the venture capital component of the investment had fallen from $822 million in 2009 to $646 million in 2010. According to the firm, U.S. private sector funding of approximately $3.5 billion led all other nations, followed by Japan (almost $3 billion) and Germany (about $1 billion). Lux Research also reported that the amount of venture capital funding in Europe was one-fifth that of the North American level.[19]

According to an analysis by the National Bureau of Economic Research, on a PPP comparison basis, the United States led the world in 2006 in corporate R&D investments in nanotechnology with an estimated $1.9 billion investment, followed by Japan with $1.7 billion. In total, U.S.- and Japan-based companies accounted for nearly three-fourths of global corporate investment in nanotechnology R&D in 2006. China ranked fifth in corporate investment, accounting for approximately 3% of global private nanotechnology R&D investments.[20]

Scientific Papers

The quantity of peer-reviewed scientific papers is considered by some to be an indicator of a nation's scientific leadership. A study by the National Bureau of Economic Research in 2005 reported that the U.S. share was a world-leading 24%, but that this represented a decline from approximately 40% in the early 1990s, concluding:

> Taken as a whole these data confirm that the strength and depth of the American science base points to the United States being the dominant player in nanotechnology for some time to come, while the United States also faces significant and increasing international competition.[21]

Reflecting the same trend, the number of papers in the Science Citation Index (SCI)[22] related to nanotechnology discoveries rose from 18,085 in 2000 to approximately 65,000 in 2008, a compound annual growth rate (CAGR) of 17.3%. The U.S. share of these papers grew at a somewhat slower pace (13.8%

CAGR) from 5,342 in 2000 to approximately 15,000 in 2008, reducing the total U.S. share from 29.5% in 2000 to approximately 23.1% in 2008.[23]

One measure of the importance of a scientific paper is the number of times it is cited in other papers. An analysis by Evaluametrics, Ltd. reports that nanotechnology papers attributed to the United States are much more frequently cited than those attributed to China, the nations of the European Union (EU27), and the rest of the world as a whole. This held true overall and separately in each of the four disciplines examined (biology, chemistry, engineering, and physics). The U.S. lead was particularly pronounced in biology. China fell below the world average number of citations in each discipline, as well as overall. The EU27 performed near the world average in engineering and physics, and somewhat higher in chemistry.

Patents

Patent counts—assessments of how many patents are issued to individuals or institutions of a particular country—are frequently used to assess technological competitiveness. By this measure, the U.S. competitive position appears to be strong. A 2007 U.S. Patent and Trademark Office analysis of patents in the United States and in other nations stated that U.S.-origin inventors and assignees/owners have

- the most nanotechnology-related U.S. patents by a wide margin;
- the most nanotechnology-related patent publications globally, but by a narrower margin (followed closely by Japan); and
- the most nanotechnology-related inventions that have patent publications in three or more countries, 31.7%, followed by Japan (26.9%), Germany (11.3%), Korea (6.6%), and France (3.6%).[24]

Environmental, Health, and Safety Implications

Key policy issues associated with U.S. competitiveness in nanotechnology include environmental, health, and safety (EHS) concerns; nanomanufacturing; and public understanding and attitudes. EHS concerns include both direct consequences for health, safety, and the environment, and how uncertainty about EHS implications and potential regulatory responses might affect U.S. competitiveness. One such effect might be the discouragement of investment in nanotechnology due to the possibility of regulations that might bar products from the market, impose high regulatory compliance costs, or result in product liability claims and clean-up costs.

Some of the unique properties of nanoscale materials—for example, small size, high surface areato-volume ratio—have given rise to concerns about their potential implications for health, safety, and the environment. While nanoscale particles occur naturally and as incidental by-products of other human activities (e.g., soot),[25] EHS concerns have been focused primarily on nanoscale materials that are intentionally engineered and produced.

Much of the public dialogue about risks associated with nanotechnology has focused on carbon nanotubes (CNTs) and other fullerenes (molecules formed entirely of carbon atoms in the form of a hollow sphere, ellipsoid, or tube) since they are currently being manufactured and are among the most promising nanomaterials. These concerns have been amplified by some research on the effects of CNTs on animals, and on animal and human cells. For example, researchers have reported that carbon nanotubes inhaled by mice can cause lung tissue damage;[26] that buckyballs (spherical fullerines) caused brain damage in fish;[27] and that buckyballs can accumulate within cells and potentially cause DNA damage.[28] On the other hand, some research has found CNTs and fullerenes to be non-toxic. In addition, work at Rice University's Center for Biological and Environmental Nanotechnology conducted in 2005 found cell toxicity of CNTs to be low and that toxicity can be reduced further through simple chemical changes to the CNT's surface.[29]

Among the potential EHS benefits of nanotechnology are applications that may reduce energy consumption, pollution, and greenhouse gas emissions; remediate environmental damage; cure, manage, or prevent deadly diseases; and offer new materials that protect against impacts, self-repair to prevent catastrophic failure, or change in ways that provide protection and medical aid to soldiers on the battlefield.

Potential EHS risks of nanoscale particles in humans and animals depend in part on their potential to accumulate, especially in vital organs such as the lungs and brain, that might harm or kill, and diffusion in the environment that might harm ecosystems. For example, several products on the market today contain nanoscale silver, an effective antibacterial agent. Some scientists have

raised concerns that the dispersion of nanoscale silver in the environment could kill microbes that are vital to ecosystems.

Like nanoscale silver, other nanoscale particles might produce both positive and negative effects. For example, some nanoscale particles have the potential to penetrate the blood-brain barrier, a structure that protects the brain from harmful substances in the blood. Currently, the barrier hinders the delivery of therapeutic agents to the brain.[30] The characteristics of some nanoscale materials may allow pharmaceuticals to be developed to purposefully and beneficially cross the blood-brain barrier and deliver medicine directly to the brain to treat, for example, a brain tumor. Alternatively, other nanoscale particles might unintentionally pass through this barrier and harm humans and animals.

There is widespread uncertainty about the potential EHS implications of nanotechnology. A survey of business leaders in the field of nanotechnology indicated that nearly two-thirds believe that "the risks to the public, the workforce, and the environment due to exposure to nano particles are 'not known,'" and 97% believe that it is very or somewhat important for the government to address potential health effects and environmental risks that may be associated with nanotechnology.[31]

Many stakeholders believe that concerns about potential detrimental effects of nanoscale materials and products on health, safety, and the environment—both real and perceived—must be addressed for a variety of reasons, including:

- protecting and improving human health, safety, and the environment;
- enabling accurate and efficient risk assessments, risk management, and cost-benefit trade-offs;
- creating a predictable, stable, and efficient regulatory environment that fosters investment in nanotechnology-related innovation;
- ensuring public confidence in the safety of nanotechnology research, engineering, manufacturing, and use;
- preventing the negative consequences of a problem in one application area of nanotechnology from harming the use of nanotechnology in other applications due to public fears, political interventions, or an overly broad regulatory response; and
- ensuring that society can enjoy the widespread economic and societal benefits that nanotechnology may offer.

Policy issues associated with EHS impacts of nanotechnology include magnitude, timing, foci, and management of the federal investment in EHS research; adequacy of the current regulatory structures to protect public health and the environment; and cooperation with other nations engaged in nanotechnology R&D to ensure all are doing so in a responsible manner.

Nanomanufacturing

Securing the economic benefits and societal promise of nanotechnology requires the ability to translate knowledge of nanoscience into market-ready nanotechnology products. Nanomanufacturing is the bridge connecting nanoscience and nanotechnology products. Although some nanotechnology products have already entered the market, these materials and devices have tended to require only incremental changes in manufacturing processes. Generally, they are produced in a laboratory environment in limited quantities with a high degree of labor intensity, high variability, and high costs. To make their way into safe, reliable, effective, and affordable commercial-scale production in a factory environment may require the development of new and unique technologies, tools, instruments, measurement science, and standards for nanomanufacturing.

Public Attitudes and Understanding

What the American people know about nanotechnology and the attitudes that they have toward it may affect the environment for research and development (especially support for public R&D funding), regulation, market acceptance of products incorporating nanotechnology, and, perhaps, the ability of nanotechnology to weather a negative event such as an accident or spill.

In 2007, the Woodrow Wilson International Center for Scholars' Project on Emerging Nanotechnologies (PEN) reported results of a nationwide poll of adults that found more than 42% had "heard nothing at all" about nanotechnology, while only 6% said they had "heard a lot." In addition, more than half of those surveyed felt they could not assess the relative value of nanotechnology's risks and benefits. Among those most likely to believe that benefits outweigh risks were those earning more than $75,000 per year, men, people who had previously heard "some" or "a lot" about nanotechnology, and those between the ages of 35 and 64. Alternatively, among those most likely to believe that the risks of nanotechnology outweigh benefits include people earning $30,000 or less; those with a high school diploma or less; women; racial and ethnic minorities; and those between the ages of 18 and 34 or over age 65.[32]

The PEN survey found a strong positive correlation between familiarity with and awareness of nanotechnology and perceptions that benefits will outweigh risks. However, the survey data also indicate that communicating with the public about nanotechnology in the absence of clear, definitive answers to EHS questions could create a higher level of uncertainty, discomfort, and opposition.

Congress expressed its belief in the importance of public engagement in the 21st Century Nanotechnology Research and Development Act of 2003 (15 U.S.C. §§7501 et seq.). The act calls for public input and outreach to be integrated into the NNI's efforts. The NNI has sought to foster public understanding through a variety of mechanisms, including written products, speaking engagements, a web-based information portal (nano.gov), informal education, and efforts to establish dialogues with stakeholders and the general public. The NSET subcommittee has also established a Nanotechnology Public Engagement and Communications working group to develop approaches by which the NNI can communicate more effectively with the public.

End Notes

[1] For additional information on these issues, see CRS Report RL34401, The National Nanotechnology Initiative: Overview, Reauthorization, and Appropriations Issues, CRS Report RL34493, Nanotechnology and U.S. Competitiveness: Issues and Options, and CRS Report RL34614, Nanotechnology and Environmental, Health, and Safety: Issues for Consideration, all by John F. Sargent, and CRS Report RL34332, Engineered Nanoscale Materials and Derivative Products: Regulatory Challenges, by Linda-Jo Schierow.

[2] National Cancer Institute website, Nanoshells, http://nano.cancer.gov/learn/understanding/nanotech_nanoshells.asp.

[3] National Cancer Institute website, http://nano.cancer.gov/resource_center/tech_ backgrounder.asp Nanotech News, Nanoparticles Enhance Combination Chemotherapy and Radiation Therapy, April 2012, http://nano.cancer.gov/action/ news/2012/apr/nanotech_news_2012-4-2f.asp; Nanotech News, First-Of-Its-Kind Self-Assembled Nanoparticle for Targeted and Triggered Thermo-Chemotherapy, December 2012, http://nano.cancer.gov/action/news/2012/dec/ nanotech_news_2012-12-13b.asp; and National Cancer Institute, NCI Alliance for Nanotechnology in Cancer, 2011 NCI Alliance Annual Bulletin, Joe Alper, Nanoparticles Deliver Drug Cocktails to Tumor, 2011.

[4] Nanoscience Research for Energy Needs, Nanoscale Science, Engineering, and Technology Subcommittee, National Science and Technology Council, The White House, December 2004.

[5] Abraham, M. "Today's Seawater is Tomorrow's Drinking Water," University of California at Los Angeles, November 6, 2006; and NNI website, Benefits and Applications, http://www.nano.gov/you/nanotechnology-benefits.

[6] Hewlett-Packard Development Company, L.P. Nanotechnology, http://www.hpl.hp.com/research/about/ nanotechnology.html; and IBM Research, Silicon Integrated Nanophotonics, http://researcher.ibm.com/researcher/ view_project.php?id=2757.

[7] Nanoscale Science and Engineering for Agriculture and Food Systems, draft report on the National Planning Workshop, submitted to the Cooperative State Research, Education, and Extension Service of the U.S. Department of Agriculture, July 2003.

[8] Kole, Chittaranjan, Kole, Phullara, et al., "Nanobiotechnology Can Boost Crop Production and Quality: First Evidence from Increased Plant Biomass, Fruit Yield and Phytomedicine Content in Bitter Melon," BMC Biotechnology, PubMed, April 26, 2013, http://www.ncbi. nlm.nih.gov/pubmed/23622112?dopt=Abstract&holding= f1000,f1000m,isrctn; and Wolfe, Josh. "Safer and Guilt-Free Nano Foods," Forbes.com, August 10, 2005.

[9] White, Scott R. and Geubelle, Philippe H., "Self-Healing Materials: Get Ready for Repair-and-Go," Nature Nanotechnology, Vol. 5, pp. 247-248, 2010, http://www.nature.com/nnano/ journal/v5/n4/abs/nnano.2010.66.html; Berger, Michael. "Nanomaterial Heal Thyself," Nanowerk Spotlight, June 13, 2007, http://www.nanowerk.com/ spotlight/spotid=2067.php.

[10] "Nanotechnology for Sensors and Sensors for Nanotechnology," Nanotechnology Signature Initiative, National Science and Technology Council, July 9, 2012, http://www.nano.gov/ sites/default/files/pub_resource/ sensors_nsi_2012_07_09_final_for_web.pdf.

[11] EPA website. http://epa.gov/ncer/nano/research/nano_remediation.html.

[12] Mihail C. Roco, "The Long View of Nanotechnology Development: The National Nanotechnology Initiative at 10 Years," Journal of Nanoparticle Research, February 2011, p. 428.

[13] For additional information on the reauthorization efforts, see CRS Report RL34401, The National Nanotechnology Initiative: Overview, Reauthorization, and Appropriations Issues, by John F. Sargent Jr.

[14] "Ranking Member Johnson Circulates America Competes Discussion Draft," House Democrats, Committee on Science, Space, and Technology, October 29, 2013, http:// democrats.science.house.gov/press-release/ranking-memberjohnson-circulates-america-competes-discussion-draft.

[15] NSET subcommittee members include Bureau of Industry and Security, DOC; Consumer Product Safety Commission; Cooperative State Research, Education, and Extension Service, Department of Agriculture (USDA); Department of Defense (DOD); Department of Education; DOE; Department of Homeland Security; Department of Justice; Department of Labor; Department of State; Department of Transportation; Department of the Treasury; EPA; Food and Drug Administration; Forest Service, USDA; Intelligence Technology Innovation Center; International Trade Commission; NASA; National Institutes of Health (NIH), Department of Health and Human Services (DHHS); National Institute for Occupational Safety and Health, Centers for Disease Control and Prevention, (DHHS); NIST, DOC; NSF; Nuclear Regulatory Commission; U.S. Geological Survey, Department of the Interior; and U.S. Patent and Trademark Office, DOC.

[16] The original six agencies were the NSF, DOD, DOE, NIST, NASA, and NIH.

[17] OECD /NNI International Symposium on Assessing the Economic Impact of Nanotechnology, Background Paper 2: Finance and Investor Models in Nanotechnology, Working Party on Nanotechnology, Organization for Economic Cooperation and Development, March 16, 2012, p. 4.

[18] Global Funding of Nanotechnologies and Its Impact, Cientifica, July 2011, available at http://cientifica.eu/blog/wpcontent/uploads/downloads/2011/07/Global-Nanotechnology-Funding-Report-2011.pdf.

[19] OECD /NNI International Symposium on Assessing the Economic Impact of Nanotechnology, Background Paper 2: Finance and Investor Models in Nanotechnology, Working Party on

Nanotechnology, Organization for Economic Cooperation and Development, March 16, 2012, p. 4.

[20] Profiting from International Nanotechnology, Lux Research, Inc., December 2006.

[21] Zucker, L.G. and M.R. Darby. "Socio-Economic Impact of Nanoscale Science: Initial Results and Nanobank," National Bureau of Economic Research, March 2005.

[22] The Science Citation Index, a product of Thomson Reuters Corporation, provides bibliographic and citation information from more than 3,700 scientific and technical journals published around the world.

[23] Mihail C. Roco, "The long view of nanotechnology development:the National Nanotechnology Initiative at 10 years," Journal of Nanoparticle Research, February 2011, p. 429. Growth rates and U.S. percentages of total publications calculated by CRS.

[24] Eloshway, Charles. "Nanotechnology Related Issues at the U.S. Patent and Trademark Office," Workshop on Intellectual Property Rights in Nanotechnology: Lessons from Experiences Worldwide, Brussels, Belgium, April 2007.

[25] Some naturally occurring nanoparticles cause adverse health effects. Studies on the effects of naturally occurring particles are numerous and inform R&D on engineered nanoparticles.

[26] Lam, C.; James, J.T.; McCluskey, R.; and Hunter, R. "Pulmonary toxicity of single-wall carbon nanotubes in mice 7 and 90 days after intratracheal instillation," Toxicological Sciences, September 2003. Vol 77. No. 1. pp 126-134.

[27] Oberdörster, Eva. "Manufactured Nanomaterials (Fullerenes, C60) Induce Oxidative Stress in the Brain of Juvenile Largemouth Bass," Environmental Health Perspectives, July 2004. Vol. 112. No. 10.

[28] "Understanding Potential Toxic Effects of Carbon-Based Nanomaterials," Nanotech News, National Cancer Institute Alliance for Nanotechnology in Cancer, July 10, 2006.

[29] "Modifications render carbon nanotubes nontoxic," press release, Rice University, October 2005.

[30] "Blood-Brain Barrier Breached by New Therapeutic Strategy," press release, National Institutes of Health, June 2007.

[31] "Survey of U.S. Nanotechnology Executives," Small Times Magazine and the Center for Economic and Civic Opinion at the University of Massachusetts-Lowell, Fall 2006.

[32] "Awareness of and Attitudes Toward Nanotechnology and Federal Regulatory Agencies: A Report of Findings," survey by Peter D. Hart Research Associates, Inc., for the Project on Emerging Nanotechnologies, September 2007.

In: Nanomanufacturing
Editor: Brad D. McReynolds
ISBN: 978-1-63117-639-5

Chapter 2

NANOMANUFACTURING: EMERGENCE AND IMPLICATIONS FOR U.S. COMPETITIVENESS, THE ENVIRONMENT, AND HUMAN HEALTH*

United States Government Accountability Office

WHY GAO CONVENED THIS FORUM

Nanotechnology has been defined as the control or restructuring of matter at the atomic and molecular levels in the size range of about 1–100 nanometers (nm); 100 nm is about 1/1000th the width of a hair.

The U.S. National Nanotechnology Initiative (NNI), begun in 2001 and focusing primarily on R&D, represents a cumulative investment of almost $20 billion, including the request for fiscal year 2014. As research continues and other nations increasingly invest in R&D, nanotechnology is moving from the laboratory to commercial markets, mass manufacturing, and the global marketplace—a trend with potential future import that some compare to history's introduction of technologies with major economic and societal impact, such as plastics and even electricity. Today, burgeoning markets, innovation systems, and nanomanufacturing activities are increasingly competitive in a global context—and the potential EHS effects of nanomanufacturing remain largely unknown.

* This is an edited, reformatted and augmented version of a report to Congressional requesters, GAO-14-181SP, dated January 2014.

At the July 2013 forum, participants from industry, government, and academia discussed the future of nanomanufacturing; investments in nanotechnology R&D and challenges to U.S. competitiveness; ways to enhance U.S. competitiveness; and EHS concerns. Participants reviewed a summary of forum discussions, and two experts (who did not attend the forum) independently reviewed a draft of this report. Their comments were incorporated in this report as appropriate.

WHAT THE PARTICIPANTS SAID

The forum's participants described nanomanufacturing as a future megatrend that will potentially match or surpass the digital revolution's effect on society and the economy. They anticipated further scientific breakthroughs that will fuel new engineering developments; continued movement into the manufacturing sector; and more intense international competition.

Although limited data on international investments made comparisons difficult, participants viewed the U.S. as likely leading in nanotechnology research and development (R&D) today. At the same time, they identified several challenges to U.S. competitiveness in nanomanufacturing, such as inadequate U.S. participation and leadership in international standard setting; the lack of a national vision for a U.S. nanomanufacturing capability; some competitor nations' aggressive actions and potential investments; and funding or investment gaps in the United States (illustrated in the figure, below), which may hamper U.S. innovators' attempts to transition nanotechnology from R&D to full-scale manufacturing.

Funding/Investment Gap in the Manufacturing-Innovation Process

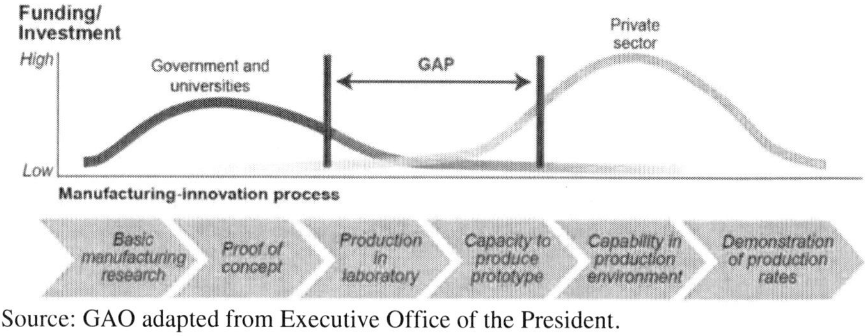

Source: GAO adapted from Executive Office of the President.

Participants outlined three approaches that might be viewed as alternative ways to address these challenges—or used together: (1) strengthen U.S. innovation by updating current innovation-related policies and programs, (2) promote U.S. innovation in manufacturing through public-private partnerships, and (3) design a strategy for attaining a holistic vision for U.S. nanomanufacturing. Participants who represented a range of perspectives on environmental, health, and safety (EHS) issues also noted that significant research is needed to understand the risks associated with nanomaterials. As such, multiple participants advocated a collaborative effort, in which nanotechnology stakeholders create an EHS framework, including developing standards for measurement and nomenclature, to help assess and address these risks.

Finally, participants advocated both maintaining R&D support and considering ways to address the challenges outlined above. Justification of further steps might be based on their potential for improving (1) international data on nanotechnology investments, (2) international standard setting for nanomanufacturing and U.S. participation, (3) U.S. ability to maintain or enhance competitiveness, and (4) U.S. and international efforts to address EHS issues.

ABBREVIATIONS

ARM	Applied Research and Methods (GAO)
ARRA	American Recovery and Reinvestment Act
BLS	Bureau of Labor Statistics
CNSE	College of Nanoscale Science and Engineering (State University of New York, Albany)
DARPA	Defense Advanced Research Projects Agency
EHS	environmental, health, and safety (issues or implications)
EIT	European Institute of Innovation and Technology (of the European Commission)
EMA	European Medicines Agency
EV	electric vehicle
FDA	Food and Drug Administration
GPT	general purpose technology
IP	intellectual property
Li-ion	lithium-ion (batteries)
MDI	manufacturing, design, and innovation

MEP	Manufacturing Extension Partnership
NASCENT	Center for Nanomanufacturing Systems for Mobile Computing and Mobile Energy Technologies (The University of Texas at Austin)
NIOSH	National Institute for Occupational Safety and Health
NIH	National Institutes of Health
NIST	National Institute of Standards and Technology
nm	nanometer (one billionth of a meter or 10^{-9} m)
NNI	National Nanotechnology Initiative
NNMI	National Network for Manufacturing Innovation
NNN	National Nanomanufacturing Network
NSF	National Science Foundation
OECD	Organisation for Economic Co-operation and Development
R&D	research and development
SBIR	Small Business Innovation Research
SIA	Semiconductor Industry Association
STEM	science, technology, engineering, and mathematics
TIP	Technology Innovation Program
TTIP	Transatlantic Trade and Investment Partnership
VC	venture capital

January 31, 2014

The Honorable Lamar Smith
Chairman
Committee on Science, Space, and Technology
House of Representatives

The Honorable Ralph Hall
House of Representatives

This report responds to your request that we conduct a Comptroller General forum on specific issues in nanotechnology manufacturing. As agreed with your staff, we designed the July 2013 forum to elicit the views of a group of leading experts on:

- nanomanufacturing's future;
- U.S. investments and competitiveness in nanotechnology research and development, as well as challenges to U.S. competitiveness in nanomanufacturing;
- ways to enhance U.S. competitiveness in nanomanufacturing; and
- environmental, health, and safety implications of nanomanufacturing.

We held the forum with assistance from the National Academies. The agenda (contained in app. I) structured forum discussions to allow each participant to comment openly and interact with others on any issue without the need to comment on all issues. We selected the forum's participants (listed in app. II) to represent (1) a range of backgrounds, covering academia, government, industry, and nonprofit organizations, and (2) experience across varied areas such as conducting nanotechnology research, developing new approaches to nanomanufacturing, commercializing and manufacturing nano-enabled products, and assessing nanomanufacturing's effect on society and the economy.

Developed under the leadership of our Chief Scientist, this report summarizes the ideas and themes that emerged in the forum's discussions. The report does not necessarily represent the views of any individual participant or organization. Similarly, the report does not necessarily represent GAO's views.

Appendixes to the report (1) list the experts we consulted, in addition to forum participants (app. III); (2) reprint forward-looking, descriptive profiles (app. IV) of four nanomanufacturing industries, which we sent to all participants in advance of the July meetings; and (3) explain the scope and methodology of our work (app. V).

We thank all the forum's participants for taking the time to share their knowledge and insights and explore the topic's implications for the United States.

Gene L. Dodaro
Comptroller General of the United States

FOREWORD

The entry of nanotechnology into manufacturing has been compared to the advent of earlier technologies that have profoundly affected modern societies, such as plastics, semiconductors, and even electricity. Applications of nanotechnology promise transformative improvements in materials performance and longevity for electronics, medicine, energy, construction, machine tools, agriculture, transportation, clothing, and other areas. Many technologically sophisticated products today (for example, smartphones, tablet computers, and targeted therapeutic drugs, among many others) already benefit from nanotechnology or some innovative nano-enabled process—as do other products that are not typically conceptualized as "high tech" (such as textiles, lubricants, and athletic gear).

However, the path to greater benefits—whether economic, social, or environmental—from nanomanufactured goods and services is not yet clear. Although many view the United States as the world's premier nanotechnology research and development (R&D) nation, some are concerned about our national ability to efficiently and effectively capture value from our collective investments, whether through intellectual property development, licensing and commercialization, manufacturing goods at scale, or delivering new services. Moreover, concerns persist regarding the environmental, health, and safety implications of nano- engineered materials.

The forum we convened in July 2013 to explore these issues offered a series of facilitated discussions. (The agenda is in app. I.) This report summarizes key themes that emerged during the forum. Specifically, the report presents participants' views concerning (1) the anticipated scope of future nanomanufacturing developments and related effects; (2) U.S. investments and competitiveness in nanotechnology R&D, as compared to other leading nations—and current challenges to U.S. competitiveness in nanomanufacturing; (3) options for enhancing U.S. nanomanufacturing competitiveness; and (4) issues in addressing the environmental, health, and safety implications of nanomanufacturing. Additionally, we discuss "considerations going forward" based on the overall forum discussions as well as participants' views on possible future actions or next steps.

We are grateful to all the participants, reviewers, and interviewees involved in formulating this report. We also acknowledge the invaluable support provided by the National Academies and the National Research Council staff.

Timothy M. Persons, Ph.D.
Chief Scientist

INTRODUCTION

Nanotechnology has been defined as the control or restructuring of matter at atomic and molecular levels in the size range of about 1 to 100 nm.[1] A nanometer (nm) is one-billionth of a meter (10^{-9} m) or about the width of an atom. To further illustrate differences in size, the width across a DNA molecule is 2 nm, the width of a red blood cell is 10,000 nm, and the width of a hair is 75,000 to 100,000 nm.[2]

Many scientific fields—such as chemistry, materials science, biology, physics, and engineering—study and apply nanotechnology; the goal is to create materials as well as devices and systems that have fundamentally new properties or functions. Worldwide there has been considerable investment, both public and private, in nanotechnology R&D. One estimate (Roco 2013) put the total worldwide figure for nanotechnology R&D investment in 2012 at approximately $24 billion.

Focusing on U.S. public investment since 2001, the overall growth in the funding of nanotechnology has been substantial, as indicated by the funding of the federal interagency National Nanotechnology Initiative (NNI), with a cumulative investment of about $18 billion for fiscal years 2001 through 2013.[3] Adding the request for fiscal year 2014 brings the total to almost $20 billion. However, the amounts budgeted in recent years have not shown an increasing trend. (See Figure 1.)

The global market for nanomaterials and nano-enabled products appears to be growing fast. According to one set of estimates (Bradley 2010), the global nanomaterials market in 2010 was expected to reach about $1.3 billion, but the market for nano-enabled products was expected to be much larger—estimated to be over $300 billion in 2010, with an expected average annual growth rate of over 40 percent for 2010 through 2014. While other estimates of market size use different definitions, estimates of growth rates in this market tend to be quite high, especially for some countries. For example, a study commissioned by the European Union (ObservatoryNANO 2012) estimated a 44 percent average annual growth rate in China's nanotechnology market between 2005 and 2010. Another issue is that some estimates and projections include certain categories of products, such as semiconductors with nanoscale features, which others exclude.

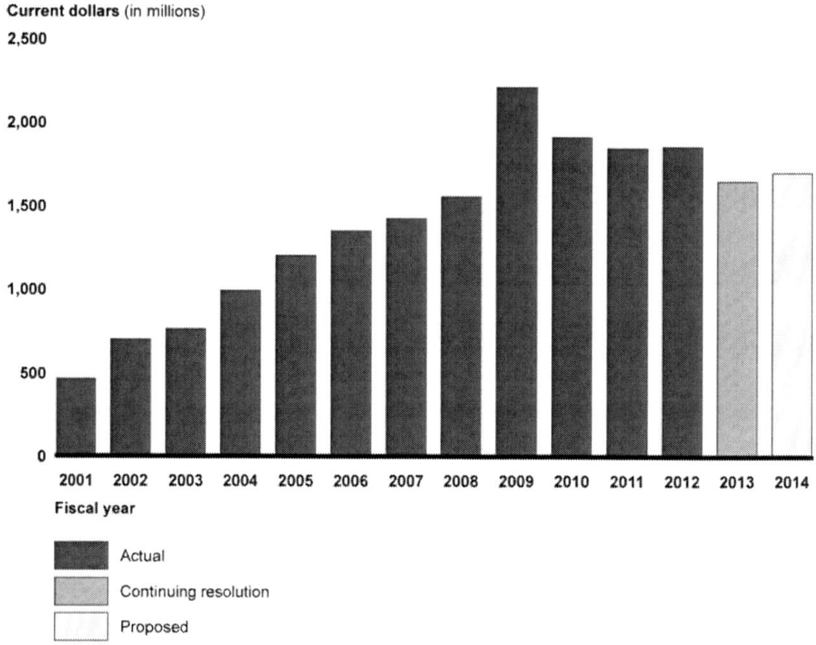

Source: GAO based on data from the U.S. National Science and Technology Council (2003; 2005 through 2013) and other sources.

Note: NNI funding is focused primarily on research and development (R&D). Amounts shown for fiscal years 2009 and 2010 include funding from the American Recovery and Reinvestment Act of 2009. The amounts budgeted for 2013 and proposed for 2014 may differ from amounts actually spent; for example, the sequestration of 2013 may have lowered amounts available.

Figure 1. U.S. National Nanotechnology Initiative Funding, Fiscal Years 2001–2014.

The use of nanomaterials in diverse consumer and commercial applications also raises questions about the potential risks that might arise if people or environments were to become exposed to nanomaterials during their manufacture, use, or disposal. Nanomaterials come in a variety of forms, based on both their chemical composition and their physical structure. The environmental, health, and safety (EHS) risks of a nanomaterial may differ by characteristics such as size, shape, and surface chemistry, among others (see Figure 2). In addition, the behavior of individuals—for example, their use or misuse of a product—can affect EHS risks. The risk posed by a material is a combination of the hazard or negative effect that it may have on an organism and the extent of the organism's exposure to it.

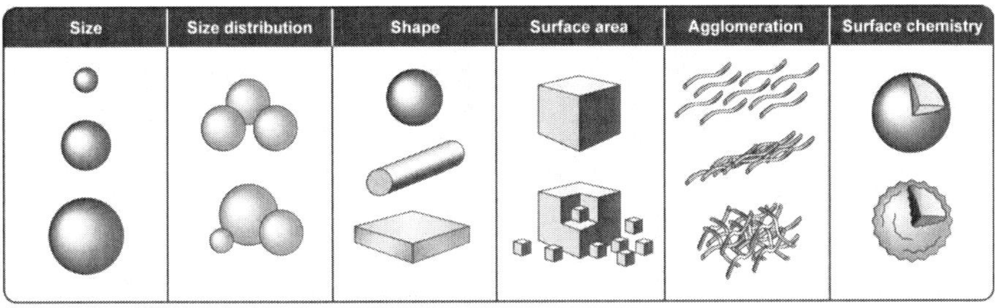

Source: GAO analysis based on GAO (2010b; 2012a), Hassellov and Kaegi (2009), and Mazov et al., (2012).

Note: Characteristics of a nanomaterial that could affect risk include its particles' (1) size, (2) distribution of sizes in a group of particles, (3) shape, (4) surface area, (5) likelihood of forming agglomerates (clumps of particles bound together), and (6) surface chemistry including surface composition, shape, or chemical reactivity.

Figure 2. Some Nanomaterial Characteristics That Could Affect Risk.

In 2012, we reported that much remains unknown about EHS risks (GAO 2012a). For example, few tools and methods exist—such as models to predict the behavior of nanomaterials in the environment—for conducting EHS research. Further, we have little information on the number of workers exposed to nanomaterials in the workplace, the extent of the release or transport of nanomaterials in the environment, or the effects on human health of such exposure. It is difficult to assess the risk of nanomaterials because they are too varied to permit generalizations about how they behave.

Nonetheless, risks associated with particular uses of specific nanomaterials can be assessed. For example, based on research studies that have shown adverse lung effects in rodents, the National Institute for Occupational Safety and Health (NIOSH) has recently released recommended exposure limits for individuals working with carbon nanotubes (NIOSH 2013a).[4] NIOSH reported that although it is not known whether humans exhibit similarly adverse health effects, the toxicity seen in short-term animal studies indicates that protective action is warranted.

Objectives

Because future developments in nanomanufacturing are likely to be affected by global economic dynamics and by continuing advances in science and technology, as well as by shifts in policy directions, the objectives of this report are to summarize forum discussions in the following areas:

- **The future of nanomanufacturing, as viewed by forum participants:** Anticipated developments concerned the future economic importance and societal impact of nanomanufacturing—for example, nanomanufacturing's prospects for (1) achieving a core capacity that can set the pace for improvements across a wide range of industry sectors (which could result in a wide array of potential societal benefits), and (2) bringing "disruptive innovation" and, potentially, a net gain in jobs.[5]
- **U.S. investments in nanotechnology R&D and current challenges to U.S. competitiveness in nanomanufacturing:** Presentations and discussions covered (1) information on this nation's public and private investments in nanotechnology R&D[6] compared to other nations' investments and (2) the challenges to global competitiveness that our nation faces, including questions about, for example, the adequacy of capital to take U.S. innovations from the laboratory to commercial production and challenges from other nations that are now investing heavily in this area, as well as issues in intellectual property protection.[7]
- **Ways to enhance U.S. competitiveness:** Approaches discussed ranged from (1) enhancing U.S. competitiveness in innovation across the economy to (2) fielding programs supporting advanced manufacturing in the United States and (3) developing a U.S. strategy for reaching a vision for nanomanufacturing.
- **Issues in addressing the environmental, health, and safety implications of nanomanufacturing:** Included here were discussions concerning EHS impacts of nanotechnology and approaches for managing risks related to those impacts, which may differ from those associated with other kinds of products.

Scope and Methodology

We selected forum participants from academia, industry, and government with the assistance of the National Academies. The forum was held July 23-24, 2013, with 28 expert participants. Appendix I contains the agenda.

To develop background information for forum participants, we selected four nanomanufacturing areas in which to create expert-based, forward-looking, descriptive profiles of nanotechnology and nanomanufacturing activity—corresponding to four of the nine nondefense industry areas listed by the National Nanomanufacturing Network (NNN) (see table 1)— and distributed them to participants.[8] These industry profiles, which we produced to frame and stimulate conversation at the forum, do not necessarily represent definitive or thorough statements about the areas discussed. We selected the four nanomanufacturing areas to (1) illustrate cross-industry variation in both the current pervasiveness of nanoproducts and the nanomanufacturing competitiveness of the United States in a global context, and (2) show that nanomanufacturing is relevant to industries that differ in terms of other factors, such as general levels of technological sophistication and rates of change.

Table 1. Four Nanomanufacturing Areas Selected for Expert-Based, Forward-Looking Profiles

Profile	NNN industry area[a]	Industry characteristics
Semiconductors	Electronics and semiconductors	Extremely fast changing industry; already pervasive products with nanoscale features. U.S. leads in design and does some manufacturing.
Battery-powered vehicles	Energy and power	Medium-to-fast rate of change; initial nano-batteries just entering the market. Very limited U.S. involvement in manufacturing these batteries.
Nano-based concrete	Materials and chemical industries	Slow-changing industry; relatively few nanoproducts in widespread use. U.S. active in related industries.
Nanotherapeutics	Pharmaceuticals, biomedical, and biotechnology	Fast changing industry; few nanoproducts in use. U.S. currently dominant.

Source: Internano.org and GAO analysis.

[a] Altogether, as noted in text, nine areas are listed by the National Nanomanufacturing Network (NNN); see http://www.internano.org/content/view/200/224/ (accessed November 20, 2013). NNN industry areas that we did not select for profiling include (1) information technology and telecommunications; (2) aerospace and automotive; (3) forest and paper products; (4) environment, infrastructure, and national security; and (5) clothing, textiles, and personal care. However, our profiles may overlap some of these areas.

Following the July forum, we began preparing for this report by providing the forum's participants (in mid-August) an initial summary of the forum discussions. We received responses from about 90 percent of participants. Almost all who responded said that they agreed with our summary, overall, and provided several clarifications of key points made during the meetings. After incorporating comments from participants, we also obtained reviews from two experts who had not participated in the meetings. Many of the forum's participants and others we consulted in preparing this report are active in nanotechnology research or manufacturing. Thus, we recognize that the forum discussions and other materials may reflect views of persons who would advocate for or might benefit from increased government funding or other supportive efforts. Therefore, we developed the scope and methodology for this report with an emphasis on presenting a carefully balanced view of the issues. Please see appendix V for a full description of our scope and methodology.

We conducted our work from November 2012 through January 2014 in accordance with all sections of GAO's Quality Assurance Framework that are relevant to our objectives. The framework requires that we plan and perform the engagement to obtain sufficient, appropriate evidence to meet our stated objectives and to discuss any limitations in our work. We believe that the information and data obtained and that the analysis conducted, provide a reasonable basis for any findings and conclusions.

PARTICIPANTS VIEW NANOMANUFACTURING AS A FUTURE MEGATREND

Forum participants described nanomanufacturing as an emerging set of developments that will become a global megatrend: a technological revolution that is now in its formative phases but that many knowledgeable persons—in science, business, and government, worldwide—expect to burgeon in the years ahead, bringing new opportunities, "disruptive innovation," jobs creation, and diverse societal benefits.[9] Multiple participants anticipated that nanomanufacturing could eventually match or outstrip the digital revolution in terms of economic importance and societal impact, as illustrated in figure 3.

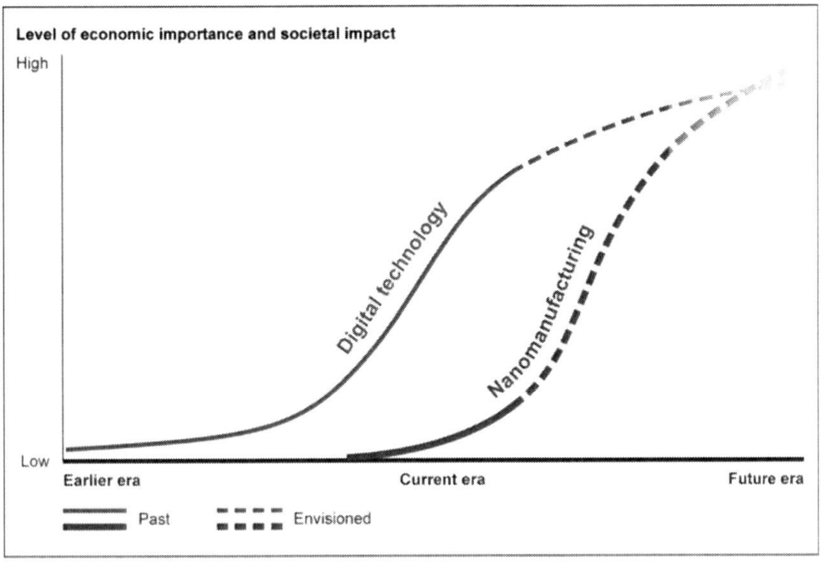

Source: GAO conceptualization based on participants' statement and the cumulative diffusion of innovation curve suggested by Rogers (1962).

Note: The envisioned "flattening" of the digital technology trend follows the "diffusion of innovation curve" (Rogers 1962). Although not shown here, the two trends may interact; that is, each may influence the other, potentially heightening one or both curves; for example, carbon nanotubes are being explored as a basis for new, smaller, and more powerful transistors (Shulaker et al., 2013).

Figure 3. Conceptualization of Nanomanufacturing and Digital Technology as Megatrends, Based on Statements of Some Forum Participants.

This bold conceptualization is linked to a variety of anticipated developments voiced by numerous forum participants. With respect to the future trajectory of nanomanufacturing—that is, the height of the nanomanufacturing trend line in figure 3, above, relative to its currently much lower level—participants said that:

- continuing breakthroughs in the science of nanomaterials are still happening and will continue to fuel engineering developments;
- enabling tools and engineering technologies, which are now in their infancy, will be replaced by new and more widely available versions that may generate significant new potential for developing new manufacturing applications; and

- the currently increasing ability of scientists and engineers to design and control nanoparticles and devices to achieve desired outcomes (such as specific product characteristics or capabilities) will speed innovation and bring major advances relative to earlier approaches, which were based on discovering a certain kind of nanoparticle and then trying to find a useful application for it.

Varied participants also (1) anticipated that by 2030 new developments may be achieved through converging technologies—for example, by combining nanotechnology with new developments in digital technology and biological science, and going forward, the convergence of nanotechnology with other new fields or technologies;[10] (2) predicted that by 2015 to 2020, molecular and hybrid nanosystems will proceed to industrial prototyping and commercialization; and (3) foresaw the possibility that nanotechnology developments might inspire new national investments. To illustrate the last point, multiple participants anticipated that conducting fundamental nanotechnology research on concrete could yield transformational advances (beyond the current level of nano-improvements in this area), which might then jump-start major federal investments in infrastructure renewal throughout the United States.

One participant saw the potential future significance of nanomanufacturing as approaching that of electricity, which had transforming impacts as it spread during the late 19th century and through the 20th century.

Multiple participants referred to nanomanufacturing as affecting many sectors of the economy and having widely transformative impacts—that is, essentially having the potential to become a "general purpose technology" (GPT), a term that has been used in reference to major innovations, such as electricity, computers, and the Internet. Innovations such as these have (1) many direct uses in numerous industries and (2) additional, indirect "spillover effects."[11] GPTs significantly affect or benefit entire national or global economies and can transform societies. (Others have specifically indicated that nanotechnology has the potential to qualify as a GPT.[12])

With respect to the potential worldwide opportunities that nanotechnology might bring, including opportunities for economic and societal benefits, various participants said that:

- *Nanomanufacturing will increasingly allow mass reproducibility at an extremely precise scale.* An historical example suggests, by analogy, the potential for nanomanufacturing's transformational effects: One hundred years ago, Henry Ford's introduction of a moving automobile assembly line and his use of interchangeable parts (which was relatively new to the automobile industry at that time) achieved a (then) new level of reproducible precision in mass production.

- *Nanomanufacturing is increasingly setting the pace for improvements across a wide range of industrial sectors with potential for an array of important societal benefits.* One participant's view of the future was that, in effect, "everything will become nano," and the result will be numerous and varied societal benefits.[13] This is already happening to some degree, based on the examples of emerging developments outlined in figure 4, below. Included here, among others, are (1) enhanced battery-powered vehicles that in the future may increasingly replace gasoline-powered models and thus help protect the environment, and (2) chemotherapy designed to target cancer cells and thus help millions of future patients avoid pain and suffering—as well as possibly saving millions of lives. New advances in applied research areas related to these examples are still occurring; to illustrate, a recent article (Yadav et al., 2013) described research on a bone-enhancing therapy that would be carried directly to the small cracks associated with osteoporosis.
- *Nanomanufacturing could eventually bring disruptive innovation and the creation of new jobs—at least for the nations that are able to compete globally.* According to the model suggested by Christensen (2012a; 2012b), which was cited by a forum participant, the widespread disruption of existing industries (and their supply chains) can occur together with the generation of broader markets, which can lead to net job creation, primarily for nations that bring the disruptive technology to market. The Ford automobile plant (with its dramatic changes in the efficient assembly of vehicles) again provides an historical example: mass-produced automobiles made cheaply enough—through economies of scale—were sold to vast numbers of consumers, replacing horse and buggy transportation and creating jobs to (1) manufacture large numbers of cars and develop the supply chain; (2) retail new cars; and (3) service them. The introduction of minicomputers and then personal computers in the 1980s and 1990s provides another historical example; the smaller computers disrupted the dominant mainframe computing industry (Christensen et al., 2000). Personal computers were provided to millions of homes, and

an analyst in the Bureau of Labor Statistics (Freeman 1996) documented the creation of jobs in related areas such as selling home computers and software. According to Christensen (2012b), "[A]lmost all net growth in jobs in America has been created by companies that were empowering—companies that made complicated things affordable and accessible so that more people could own them and use them."[14] As a counterpoint, a recent report analyzing manufacturing today (Manyika et al., 2012, 4) claims that manufacturing "cannot be expected to create mass employment in advanced economies on the scale that it did decades ago."

- *Nanomanufacturing could open new world markets.* One participant suggested that a future direction in nanomanufacturing might bring the production of very low cost goods similar in function to existing products; such goods might open untapped markets at the "bottom of the pyramid" and afford vast numbers of people new access to them. Logically, economic benefits and jobs would flow to those able to produce the new minimal cost nanoproducts. (Potentially, spin-off jobs might also be created in the consuming nations, but many "old jobs" might also be lost.)

Given the potential opportunities now anticipated, and considering that economic security is related to national security, participants also foresaw intense nanotechnology competition on a global scale—with one participant saying that we are already "in a moon race."

Stage 1: Evolving nanoscale materials, components, or devices	**Stage 2:** Nano-enabled products or intermediates	**Stage 3:** Improved or new nano-enhanced products
Nano-transistors	further improved semiconductor chips	ever faster computers, smaller smart phones
Copper nano-wires	lithium-ion batteries	more powerful battery-powered vehicles
Carbon nanotubes	concrete additives that conduct electricity	road pavement with remote sensing
Protein nanoparticles	carriers of chemo-therapeutic drugs	chemotherapy targeted to cancer cells (only)

Source: Forum presentation (Persons 2013).

Notes on value chains, "ever faster computers," and copper nano-wires: (1) We define a value chain, for purposes of this report, as a series of key steps starting with the processing of raw materials and continuing to the production of a finished consumer product; each step adds value—and may or may not involve a different company or intermediate product. The figure uses three main stages to summarize four examples of nanotechnology value chains. (2) With respect to "ever faster computers," digital development has generally followed "Moore's law," in part by utilizing chips with nano-features; however, more advanced applications of nanotechnology—such as the use of new, more advanced nanomaterials in conjunction with 3D chip architecture, optical interconnects, or other approaches—may be needed for continuous improvement in future decades. (3) The use of copper nano-wires is one example of how nanotechnology can be used to enhance lithium-ion (Li-ion) batteries, as illustrated in Figure 10 of this report.

Figure 4. Diverse Value Chains Involving Nanoscale Materials, Components, or Devices, as of 2013.

Participants See the U.S. as Likely Leading in R&D—but Facing Challenges to Competitiveness

The United States has made significant investments in nanotechnology R&D, and while recognizing the uncertainty associated with cross-nation comparisons, forum participants viewed the United States as, overall, likely leading in nanotechnology R&D at the present time.[15] This view is consistent with (1) an apparent U.S. lead in R&D investment, based on publicly available information, and (2) U.S. dominance in publications in three highly cited journals.

Participants also viewed the United States as facing several challenges that, taken together, could represent a significant obstacle to U.S. competitiveness in nanomanufacturing, going forward. These challenges include:

- *Gaps in U.S. nano-commercialization funding or investment and related issues, which may hamper nano-innovation in the United States.* Participants described key gaps in government funding and private-sector investment for not only technology development but also manufacturing development—gaps that they referred to as the *Valley of Death* and the *Missing Middle*, as explained below.

- *Prior U.S. "offshoring" of manufacturing and, in some cases, current workforce limitations.* Multiple participants said that such factors can translate to competitive disadvantages when efforts are made to establish and maintain nanomanufacturing in the United States.[16]
- *Lack of sufficient U.S. participation in setting standards for nanotechnology or nanomanufacturing.* Some participants discussed a possible need for a stronger role for the United States in setting commercial standards for nanomanufactured goods (including defining basic terminology in order to sell products in global markets).[17]
- *Lack of a national vision for a U.S. nanomanufacturing capability and limited technology transfer at U.S. universities.* Varied participants felt that such issues could make it more difficult to translate R&D investments in nanotechnology to commercial products—and thus could represent important limitations.
- *Global competition and threats.* Participants said that in key competitor countries, nano-commercialization funding or investment gaps (which present barriers to some U.S. innovators) do not apply or are being addressed. They also said that some other countries, or certain elements in other countries, are "playing by new rules"—using tactics that can threaten U.S. competitiveness in a globalized economy.

Despite Some Uncertainty, Participants See the U.S. as Likely Leading in Nanotechnology R&D

Despite some uncertainty, forum participants view the United States as likely leading in nanotechnology R&D at the present time. Specifically, participants see the United States as (1) appearing to lead in nanotechnology R&D investments, based on publicly available data, and (2) dominating in quality research. However, one participant—who agreed that the United States likely leads in nanotechnology R&D—said that this lead is increasingly being challenged in various areas by other countries.

Participant presentations included information on:

- investments in nanotechology R&D including (1) public-sector investments made by the United States and other individual nations, and (2) combined public- and private-sector investments made by the United States and the world as a whole; and

- indicators of leading nations' competitiveness in nanotechnology R&D, such as numbers of publications in leading journals.

U.S. and Other Nations' Investments in R&D

Both public-sector and private-sector investments in R&D are relevant. First, with respect to public-sector investments, some leading countries (such as the United States, some European countries, and Japan) generally track their governments' expenditures on nanotechnology R&D and make these estimates available to the public. Importantly, however, participants identified two points limiting international comparisons of current public investments in this area:

- Relevant definitions—for example, how nanotechnology itself and R&D are defined—may vary across nations, thus reducing the validity of comparisons.
- Some countries that appear to have made major investments in nanotechnology R&D may (1) not adequately or effectively track these investments or (2) not share such information externally—resulting in limited information on which to base estimates or projections.

Projections of public investments from a consulting firm are available, however, and were presented by a forum participant. Figure 5 compares the public investment in R&D made by the United States in 2013 to investments made by other selected leading investor nations—based on (1) the U.S. federal budget's figure for nanotechnology R&D, (2) projections for the U.S. states and other nations, and (3) the uncertainty that two key forum participants associated with these projections.[18] In addition to considering indicators of individual nations' public investments in nanotechnology R&D as of 2013, a forum participant

- presented an R&D public investment level for Western European governments—based on combining projected investments by all EU countries' individual governments and a projected European Commission investment—which was higher than that estimated for any single nation, and

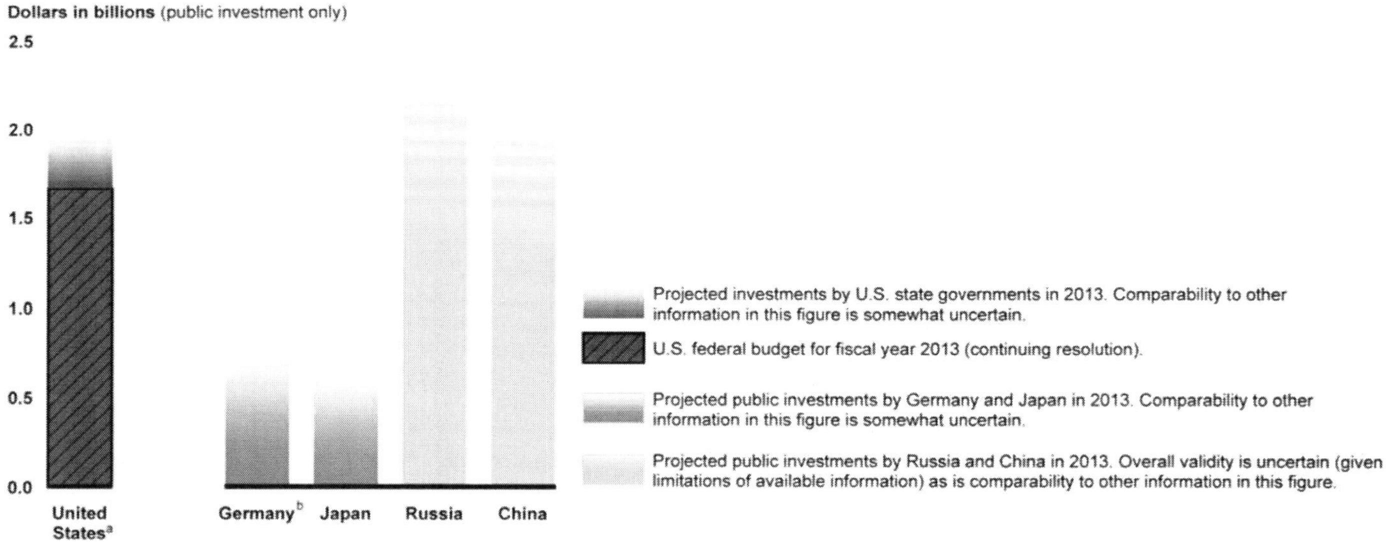

Source: GAO based on a forum presentation (Roure 2013), Cientifica, Ltd. (2013), and the U.S. National Science and Technology Council (2013).

Note: The shading in this bar chart characterizes the uncertain levels of public investment in R&D, based on two key participants' opinions concerning available projections for 2013. Specifically, the lighter the color of a bar, the greater the uncertainty associated with the projection. Use of fading on the upper portion of bars is also intended to convey the uncertainty associated with these projections. Our intent is to avoid conveying an unwarranted level of precision, which might be associated with a specific data point for each nation (see app. V on scope and methodology). Importantly, this figure excludes estimates of private-sector investments, which are discussed later in this section. Finally, this figure shows indicators of public investments for fiscal year 2013, and we note that different nations' definitions of a fiscal year may vary.

[a] Public investments shown for the United States include both state investments (projection) and the federal investment represented by the 2013 budget (continuing resolution) for the National Nanotechnology Initiative or NNI. The NNI focuses primarily on R&D. Amounts spent by NNI agencies and departments may vary from budget figures because of factors such as sequestration.

[b] The projected public investment for Germany does not include its contribution to the European Commission's effort in nanotechnology R&D.

Figure 5. Public Investments in Nanotechnology R&D in 2013—Comparison of the U.S. to Selected Leading Investor Nations, Based on (1) the U.S. Federal Budget; (2) Available Projections for Other Government Investments; and (3) the Uncertainty Associated with These Projections.

- emphasized that federal U.S. funding, as represented by the NNI budget, has not increased from 2012 to 2014.

Second, turning to private-sector investments in nanotechnology R&D,[19] uncertainty about various nations' levels of investments again derived from issues such as varying definitions and possible lack of transparency for some countries. Estimates for multiple individual nations' private- sector investments (alone) were not presented at the forum. However:

- A participant did present an estimate of *combined* public- and private-sector R&D investment in nanotechnology in 2012 for the United States—which was more than twice the figure for U.S. public-sector investment, alone (Roco 2013).
- Additionally, two participants (who were selected for the forum, in part, because of their knowledge about nanotechnology R&D investments) indicated that based on publicly available information, (1) private- sector R&D investments are significant for some countries besides the United States and that (2) the United States currently appears to lead in combined public- and private-sector investments in nanotechnology R&D, but it is possible that one or more nations might be making a greater investment than the United States (because their investments may be underreported).

Leading Nations' R&D Competitiveness

Turning to leading nations' *competitiveness in nanotechnology R&D*, a forum participant presented data from three highly cited journals, showing the percentages of articles on nanotechnology that were authored by researchers in the five top countries (defined by authorship in these journals). The United States led from 1991 to 2012. The U.S. share in 2012 was approximately 56 percent, although one participant saw this figure as likely overstating U.S. dominance.[20] Portions for the United States and the other four countries in 2012 are illustrated in figure 6.

At the same time, China overtook the United States in 2010 in terms of the *quantity* of nano-science articles published annually (a comparison made without controlling for quality of publication vehicle). However, participants generally viewed the United States as dominating in what might be considered quality research on nanotechnology.

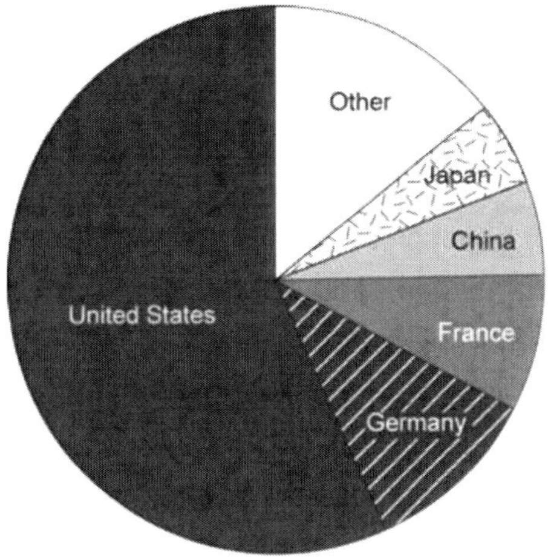

Source: GAO graphic adapted from a forum presentation (Roco 2013).

Notes: (1) Roco (2013) defines highly cited journals as *Science*, *Nature*, and the *Proceedings of the National Academy of Sciences*. (2) For articles with multiple authors, country was defined as that of the corresponding author.

Figure 6. One Participant's Indicator of Competitiveness in Nanotechnology R&D: Relative Portions of Articles in Three Highly Cited Journals for the United States and Four Other Leading Countries, 2012.

Overall Comparison of the U.S. to Other Nations: Nanotechnology R&D

Despite some uncertainty about the quality of available data on international investments, forum participants—who, in some cases, have interacted extensively with nanotechnology experts in other nations— agreed that currently the United States likely leads other nations in nanotechnology R&D, overall. Their position is consistent with (1) an apparent current U.S. lead in investments in nanotechnology R&D (combined public- and private-sector investments), and (2) U.S. competitiveness in R&D, as represented by U.S. dominance in three highly cited journals. Although forum participants saw the likely U.S. lead in nanotechnology R&D as an important strength, they foresaw intense global-scale competition in nanotechnology. Other nations are now making large investments in nanotechnology R&D—and multiple participants emphasized that it is essential for the United States to maintain a high level of investment in fundamental research, even for relatively mature industries such as the semiconductor industry, which is based on fundamental materials science. As noted in the previous section, one participant characterized this competition as a "moon race."

Gaps in Funding or Investment May Hamper U.S. Nano-Innovation

According to forum participants, nano-innovation may involve developing not only (1) a new technology or product but also (2) a new manufacturing process to produce that product. For each of these efforts, a gap in U.S. government funding, private investment, or other support may occur during the middle-stages of innovation or development. Such gaps may hamper U.S. nano-innovators as they attempt to transition from basic nanotechnology or nanomanufacturing research and laboratory-scale development to large-scale commercial nanomanufacturing.

Participants said that in the United States, government often funds research and the initial stages of development, whereas industry investment typically focuses on the final stages; however, the result is that middle-stage funding or investment for nano-innovation is insufficient. Middle-stage gaps are important because, according to participants, significant costs are associated with transitioning nano-based innovations to the prototype stage and, more so, to the scale-up manufacturing phase.

Varied participants discussed the limited federal funding available for commercialization and a "draining away" of VC for nano-innovators. Participants also mentioned the lack, in some cases, of a supportive regional ecosystem—and one participant provided an example of a U.S. nano-innovator who was able to combine varied sources of funding, including support from foundations.

Technology Innovation and Manufacturing Innovation: a Dual Challenge

Varied participants explained that a nano-innovator may face a dual challenge of technology innovation and manufacturing innovation— although these two processes may be intertwined. Using the example of nanotherapeutics for drug delivery, at the same time that innovators at the University of North Carolina advanced the maturity of specific nanotherapeutic approaches, they *also had to create ways to produce* the new nanotherapeutics: first, in the laboratory and later, in a production environment.

To illustrate this, figure 7 shows (1) an example of a specific particle-shape required for a certain type of nanoscale drug delivery and (2) the method developed to produce this and other specifically shaped particles.[21] Differently shaped and sized particles can be specially designed to maximize uptake by targeted cells and to minimize uptake by other cells. The particles are then manufactured using a process such as that shown in figure 7 (right-hand illustration).

To illustrate the extensive work that may be involved, many patents may be needed to complete development of both a new nano-enhanced product (or a series of products) and a new nanomanufacturing process. According to a forum participant involved with the start-up company that developed the PRINT® production process, shown in figure 7, this process was launched with three seminal patents in 2004. Many subsequent applications and processes were needed (approximately 80 patents pending thus far) to help the technology realize full implementation.

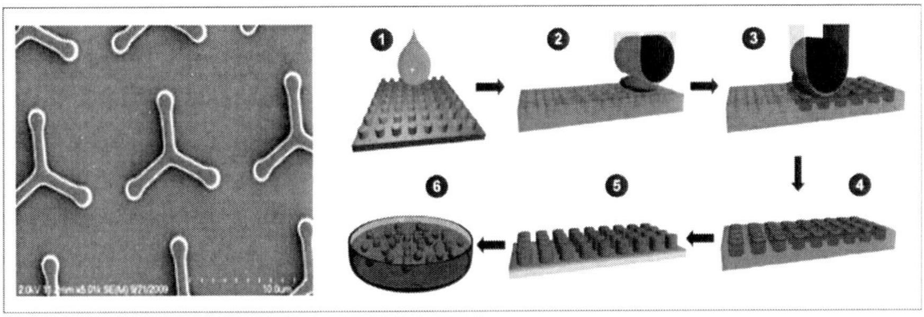

Source: University of North Carolina at Chapel Hill (DeSimone Research Group) and Liquidia Technologies.

Note: Each of the specially designed particles, shown magnified above left, is roughly one-tenth the width of a hair, or less. The production process, shown at right, begins with the nanoscale, lithographic patterning of a template, which is illustrated as a grey plate in (1) above. The template defines the size and shape of the particles to be produced. A liquid polymer illustrated as a green drop, see (1) above, is spread across the patterned template, filling the space around all the nanosize features. The polymer is then cured and becomes a solid inverse, which is used as a mold, illustrated as the green plate in (2) above. The mold is filled with a nanoparticle material, as illustrated in red; see (2), (3), and (4), above. A harvesting film, illustrated as a clear strip shown in (5) above, is used to extract the particles from the mold. Each resulting particle, illustrated in (6) above, is of the same size and shape; each also has the same chemical composition.

Figure 7. Example of Specially Designed Particles and Overview of the PRINT® Technology Process That Produced Them.

The Valley of Death, the Missing Middle, and the High Cost of Nano-Innovation

Forum participants said that middle-stage funding, investment, and support gaps occur for not only technology innovation but also manufacturing innovation. They described the *Valley of Death* (that is, the potential lack of funding or investment that may characterize the middle stages in the development of a technology or new product) and the *Missing Middle* (that is, a similar lack of adequate support for the middle stages of developing a manufacturing process or approach), as explained below.

The *Valley of Death* refers to a gap in funding or investment that can occur after research on a new technology and its initial development—for example, when the technology moves beyond tests in a controlled laboratory setting.[22] In the medical area, participants said the problem of inadequate funding/investment may be exacerbated by requirements for clinical trials. To illustrate, one participant said that $10 million to $20 million is needed to bring a new medical treatment into clinical trials, but "support from [a major pharmaceutical company] typically is not forthcoming until Phase II clinical trials," resulting in a *Valley of Death* for some U.S. medical innovations. Another participant mentioned an instance where a costly trial was required for an apparently low risk medical device—and this participant tied high costs of this type to potential difficulties that medical innovators might have obtaining venture capital. A funding/investment gap at this stage can prevent further development of a technology.

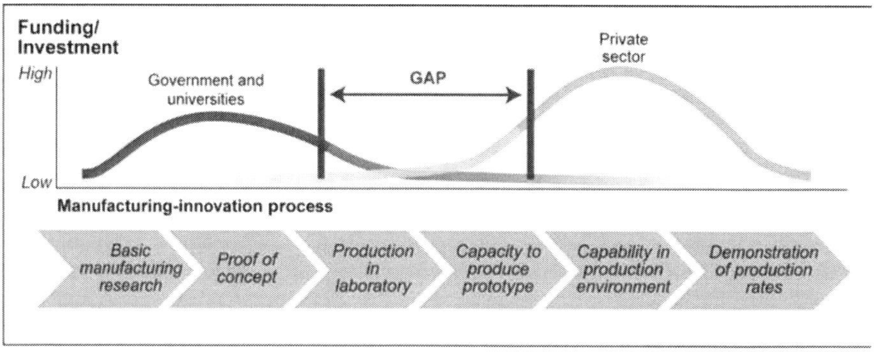

Source: GAO adapted from Executive Office of the President, 2012, page 21.

Figure 8. Funding/Investment Gap in the Manufacturing-Innovation Process: the *Missing Middle*.

The term *Missing Middle* has been used to refer to the lack of funding/investment that can occur with respect to manufacturing innovation—that is, maturing manufacturing capabilities and processes to produce technologies at scale, as illustrated in figure 8.[23] Here, another important lack of support may be the absence of what one participant called an "industrial commons" to sustain innovation within a manufacturing sector.[24] Logically, successful transitioning across the middle stages of manufacturing development is a prerequisite to achieving successful new approaches to manufacturing at scale.

Substantial amounts of funding/investment are needed to transition through the *Valley of Death* and the *Missing Middle*. High costs can act as an effective barrier to entry for small and medium-sized companies that have innovations in technology but lack the resources needed to carry their innovations all the way to commercialization and full-scale production. Further, an expert in the semiconductor industry told us that even very large companies may be unable or unwilling to risk the large investments needed to bring new, innovative technologies to market because of the exceedingly high costs involved.

Limited Federal Funding for Nano-Commercialization

Focusing on federal support for nano-commercialization, participants said this is generally limited, and as one participant explained, U.S. federal policy tends to reflect a commitment to allowing private-sector markets to identify and invest in "winning" innovations (rather than the federal government's taking this role). However, there have been significant exceptions. To illustrate:

- Multiple forum participants made positive comments about the U.S. Small Business Innovation Research (SBIR) program. Phases I and II of SBIR support R&D and establish commercialization potential (for nanotechnology as well as other areas).[25] Although the SBIR website specifically states that its policy rules out directly funding Phase III (the pursuit of the commercialization of technologies developed during Phases I and II),[26] various agencies in the SBIR program have created mechanisms or ways to indirectly support Phase III. According to a National Research Council report (Wessner 2008, 9), some agencies have created mechanisms "such as the Navy's Phase

IIB SBIR or Phase III funding with program dollars to help bridge the Valley of Death." And we subsequently reported (GAO 2011a, 13) that "NIH offered selected current or past phase II award recipients the opportunity to work one-on-one with an advisor over a 9-month period to develop business plans to commercialize their technologies ... [and] since 2004, almost 700 award recipients have received the assistance."

- Additionally, a participant pointed to the American Recovery and Reinvestment Act of 2009 (ARRA), which for example, included major funding to build a manufacturing supply chain for advanced batteries for electric vehicles.[27] Others mentioned that the Department of Defense's funding, which may extend to commercialization, can help maintain industries that enhance national security (and today, these are likely to involve nanotechnology). However, such federal investments may apply in only limited instances.

A Possible "Draining Away" of Venture Capital

Turning to private-sector funding, multiple forum participants said that in recent years, venture capital (VC) has "drained away" from relatively costly physical science areas such as nanotechnology to fund new ventures in Internet services, which generally provide larger and faster returns for a given level of investment.[28] More generally, there may be a reluctance of private sources of funding, including VC, to embark on ventures that are perceived as high risk and high cost, or that might take too long to start realizing positive cash flows or might fail altogether. A related development was noted by a survey of 270 U.S. nanomanufacturers, which a participant cited. Specifically, that survey (Mehta 2010a, 2010b) identified limited capital as a top barrier to nanomanufacturing. Thus, although one participant's presentation estimated U.S. VC investments in nanotechnology at $1 billion in 2010 (Roco 2013) and another participant noted that some large companies provide VC to support innovations that will enhance their own future businesses,[29] such investments may be far from sufficient, overall. One participant's post-forum suggestion was that there is a *lack of adequate information* on the levels of investment in nanotechnology by various sources (such as VC and major corporations) and that this area deserves inquiry.

Regional Ecosystems, Foundations, and an Example of Multi-Source Support

Participants noted that nano-innovators may lack a supportive regional ecosystem (which can, in some cases, make the whole greater than the sum of its parts). Some regions provide strong comparative advantage to developers of innovative technologies in a particular field. This advantage derives from the presence of firms engaging in similar or complementary activities, a well-trained labor force, financiers, marketers, and legal specialists—as well as a developed transportation and communication infrastructure that helps innovators overcome engineering, business, institutional, and supply-chain challenges.[30]

Two participants mentioned an alternative or addition to VC and federal funding—philanthropic support from entities such as foundations. Varied funding sources may be required to complete the innovation process. A forum participant told us, in post-forum communications, that over the past 9 years, the start-up he founded raised tens of millions of dollars from various sources. These included foundation support as well as venture capital and federal funding encompassing an SBIR grant and a National Institute of Standards and Technology (NIST) Technology Innovation Program (TIP) grant.

Prior Offshoring and Other Issues Challenge Efforts to Establish and Maintain Nanomanufacturing in the United States

Multiple participants said the challenge of establishing widespread nanomanufacturing in key industries in the United States is heightened by issues such as earlier offshoring and loss of U.S.-based industries.[31] One forum participant, speaking of advanced batteries, pointed out that when "we design here [and] ship [manufacturing] abroad, we lose this shop- floor-innovation kind of mentality"—in other words, too much offshoring of skilled manufacturing jobs can, in general, put the United States at a disadvantage in terms of its ability to maintain design expertise.[32] Other participants provided examples suggesting that earlier offshoring, loss of key industries, or loss of U.S. leadership can lead to, for example:

- a continued reliance on manufacturing facilities located overseas, even as increasingly automated processes diminish the need for low cost labor, or

- a lack of U.S. manufacturing expertise in certain areas, making it difficult for the United States to enter or re-enter those areas and compete successfully.

Three examples (semiconductors, advanced batteries for vehicles, and cement-based construction materials) illustrate, in the following sections, how earlier offshoring, loss of key industries, and loss of U.S. industry- leadership may negatively affect nanomanufacturing in the United States. Additionally, difficulties finding appropriately skilled workers may challenge U.S. nanomanufacturers, as discussed below.

Earlier Offshoring and the Semiconductor Industry

The first example concerns the semiconductor industry. Despite U.S. strength in the design of semiconductors, most (not all) fabrication of semiconductors—which includes computer chips with nanoscale features—takes place outside the United States. The actual manufacturing of semiconductors is automated (as shown in Figure 9) and engineers are the key employees at a fabrication plant.[33]

Source: College of Nanoscale Science & Engineering, State University of New York.

Figure 9. View of a Semiconductor Manufacturing Facility.

But significant amounts of unskilled labor are needed for the *packaging* of semiconductors, according to one participant.[34] That participant said he believes it is important for the United States to establish more fabrication plants in this country. With respect to the possibility of this happening in the future, we note that one company is now developing advanced robotics for semiconductor packaging—which would lessen the need for unskilled labor.[35]

Loss of a Key Industry and the Case of Nano-Enhanced Batteries for Hybrids and EVs

The second example concerns the manufacture of lithium-ion (Li ion) nano-enhanced batteries that are used to power hybrids and electric vehicles (EV). Most lithium-ion batteries—including those used to power hybrids and EVs—are manufactured in Asia, although the United States developed the underlying technology and is currently investing heavily in nano-R&D for vehicle batteries. According to one forum participant, the reason is that:

(1) Smaller lithium-ion batteries for consumer electronics, such as cell phones, have long been manufactured in Asia (because the United States "gave up on [that industry] some time ago"); and
(2) Asian firms appear to have a definite competitive advantage in the manufacturing process for lithium-ion batteries, which is similar for small and large batteries.[36]

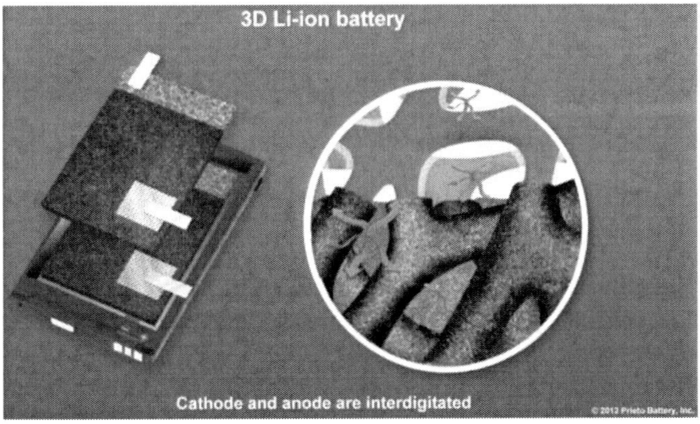

Source: Prieto Battery, Colorado, USA.

Note: The term *interdigitated*, which is used in illustration above, refers to an interlocked or interwoven design in which copper nano-wires (anode) are coated with an ultra-thin electrolyte, and then surrounded by a cathode slurry.

Figure 10. Illustration of a 3D Lithium-Ion (Li-ion) Battery.

However, looking to the future, one participant felt that "the jury is still out" on whether the United States can successfully compete in this area.[37] Another expert (whom we interviewed in preparation for the forum) said that some future versions of nano-engineered batteries will require *different manufacturing processes* and thus might represent a new opportunity for U.S. manufacturing. An example of a new type of nano- engineered battery—now in the research or idea stage with respect to use in hybrids and EVs—is shown in figure 10.

Loss of U.S. Leadership in the Cement Industry

The third example concerns the cement industry. One forum participant said that companies headquartered in other countries now control the cement industry (which was previously dominated by the United States). While cement for domestic use is produced in the United States and important university research is being pursued here, the dominant companies—headquartered elsewhere—are spearheading development of new technologies, which are thus likely to be first adopted overseas. This participant and another said that, without a U.S. industry focus on nano-concrete and without a national R&D agenda or commercialization plan in this area, the U.S. concrete industry may continue to lag—and the United States may become a late-adopting consumer of innovative technologies. One potential implication is that needed modernization of U.S. infrastructure (for example, the U.S. highway transportation system, which is a critical enabler of economic vitality) could lag behind efforts taken by other countries and might impair U.S. economic competitiveness.

Difficulty Finding Appropriately Skilled Workers

Turning to workforce issues, forum participants said that nanomanufacturing (like other forms of advanced manufacturing today) is very different than manufacturing was in the past—and requires new skills. Today's advanced manufacturing is illustrated by the view of a semiconductor-manufacturing facility shown earlier in figure 9. Changes in needed skills would seem to be especially applicable to jobs in innovative high-tech manufacturing areas, such as nanomanufacturing. Various participants said that:

- The new kinds of skills required for high-tech manufacturing jobs can be very specialized, for example, powder processing and metallurgy extraction.

- There are many manufacturing jobs that in the past required a high-school degree, but that now require an associate's degree plus aptitude for training in areas such as math and statistics.

Multiple participants said that despite current, relatively high unemployment, some high-tech manufacturers report they cannot find appropriately skilled workers; in fact, the participants said that difficulty finding sufficient numbers of skilled workers is a key reason limiting expansion of high-tech businesses.[38]

Further, multiple participants described an underlying problem—perhaps an industry communication problem—in that the general public's perception of manufacturing has not kept pace with trends in this area. One said that some young people today reject manufacturing as a career choice, based in part on misperceptions of the current nature of advanced manufacturing and the kinds of jobs that are available.

Insufficient U.S. Efforts to Set Standards May Limit Nano-Innovation

Varied participants pointed to the current lack of a unified system to describe nanomaterials—including naming conventions, definitions, and standards—as a possible limitation on innovation efforts. As one participant noted, such a system is needed for the creation of a database of nanomaterials.[39] Further, if such a system were developed, it might enhance, in the words of one participant, "the capacity to scale up innovation… [the creation of] revenue downstream ... [and] conditions for international trade [and] security—which is important [for] investors, citizens and consumer groups."

One participant noted that nanotechnology standards have been issued by American standard-setting organizations for nomenclature and measurement, but that international standard-setting efforts have been more challenging (referring to developing ISO standards).[40] Participants stressed the lack of funding and time for participating in these efforts and traveling internationally to do this.

Lack of U.S. Vision for Nanomanufacturing and Possible Inadequacy of Technology Transfer May Limit Competitiveness

Varied participants highlighted potential U.S. under-attention to two issues that may limit our nation's competitiveness, especially in nanomanufacturing, as follows:

- A forum participant said that the United States lacks a vision for a nanomanufacturing capability. Although a forum presentation mentioned four centers, funded by the National Science Foundation (NSF), that are focused on new concepts and development of methods for nanomanufacturing (that is, centers conducting research primarily at "early *manufacturing* readiness levels"[41]), our post-forum communications with an NSF official indicated that (1) the funding for three of the centers will end in 2014, and for the fourth, in 2015 and that (2) there is no program devoted to supporting nanomanufacturing centers.[42]
- Other participants, emphasizing that high-quality research and intellectual property (IP) are associated with universities across the United States, voiced the opinion that technology transfer capabilities in some universities might be improved by their adopting best practices—to increase the likelihood of research results "making it out of the university" and into industry. According to various forum participants, such U.S. universities might (1) increase investments in technology transfer; (2) assure that technology transfer staffs have sufficient business experience; and (3) increase incentives for the transfer of IP into the commercial sector. Where needed, improvements such as these might enhance commercialization in high-tech areas of endeavor (such as nanomanufacturing). However, one participant, in a post-forum communication, questioned the need for improvements in technology transfer; this participant felt that existing incentives for universities and academic researchers may be sufficient.

Global Competition and Threats to U.S. Industry

Participants said that other countries (1) are less affected by funding and investment gaps—or have begun to address them; (2) are acquiring or attempting to acquire U.S. nano-companies or nano-researchers; and (3) in some cases, are associated with threats to the intellectual property of U.S. researchers and manufacturers.

Valley of Death and Missing Middle Less Problematic in Other Nations

Participants said that the *Valley of Death* and *Missing Middle* funding and investment gaps, which are of concern in the United States, do not apply to the same extent in some other countries—for example, China and Russia—or are being addressed. One participant said that other countries in which these gaps have occurred "have zeroed in [on them] with a laser beam." Another participant summed up his view of the situation with the statement: "Government investments in establishing technology platforms, technology transfer, and commercialization are higher in other countries than in the United States." He further stated that those making higher investments include China, Russia, and the European Union.

Multiple participants referred to the European Commission's upcoming Horizon 2020 program, which will have major funding extending over 7 years. In addition to providing major funding for fundamental research, the Horizon 2020 website states that the program will help to:

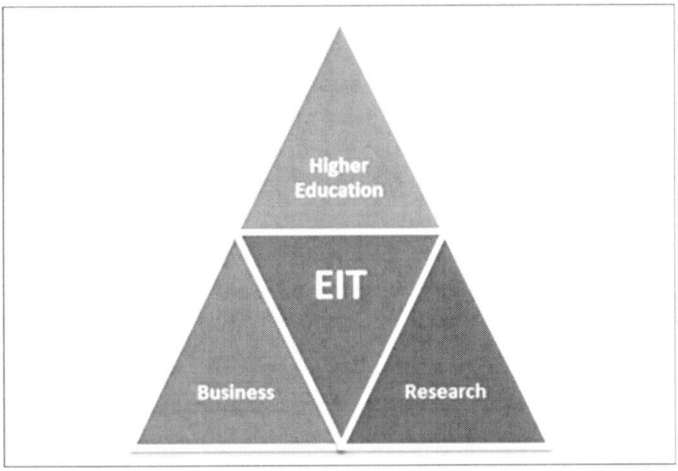

Source: "Knowledge triangle" illustrating European Commission's European Institute of Innovation and Technology principles, July 9, 2013. http://ec.europa.eu/education/eit-about_en.htm.

Figure 11. Knowledge Triangle: the Approach of the European Institute of Technology and Innovation (EIT).

> "...bridge the gap between research and the market by, for example, helping innovative enterprises to develop their technological breakthroughs into viable products with real commercial potential. This market-driven approach will include creating partnerships with the private sector and Member States to bring together the resources needed."

A key program within Horizon 2020 consists of the European Institute of Innovation and Technology (EIT), which as illustrated in the "Knowledge Triangle" shown figure 11, below, emphasizes the nexus of business, research, and higher education. The 2014-2020 budget for this portion of Horizon 2020 is 2.7 billion euros (or close to $3.7 billion in U.S. dollars as of January 2014).

Also relevant to the *Missing Middle* as this applies to U.S. firms, competition from other countries, and resulting impacts, multiple participants referred to the previously mentioned case of lithium-ion batteries as an example of the United States' having earlier successfully funded R&D, but other countries' then commercializing and mass- producing the resulting products.

Acquisition of U.S. Nano- Companies and Offers to U.S. Nano-Innovators by Other Nations

Multiple participants also said that certain other countries are purchasing struggling U.S. nanotechnology companies and making significant offers to U.S. nanoscience researchers and nanotechnology innovators alike. For example, they discussed the Chinese purchase (in early 2013) of the bankrupt A123, a nano-engineered-battery company founded by a researcher at the Massachusetts Institute of Technology. Moreover, outside of formal sessions, multiple participants described instances where other countries have made substantial offers to leading nano-researchers. Similarly, earlier congressional testimony by a university professor who is a nanotechnology researcher (Tour 2011) described multiple overtures and actual offers from other governments and companies in other countries. Such offers are not inconsistent with the statements of a nanotechnology innovator who said (in a pre-forum interview) that "going to Asia" for funding was "pervasive" in his area of work.

International Threats to U.S. Intellectual Property

Several participants discussed threats to IP associated with global competition.[43] One participant described persistent attempts by other countries (or by certain elements in other countries) to breach information systems at his nanomanufacturing company. Another described an IP challenge pertaining to research at U.S. universities, as follows:

- due to a culture of openness, especially among students, ideas and research are "leaking out" of universities prior to the initial researchers having patented or fully pursued them;
- there are many foreign students at U.S. universities; and
- there is a current lack of awareness about "leakage" and of university policies or training to counter it.

Additionally, one of our earlier interviewees said that one country targeted specific research projects at U.S. universities—and then required its own citizen-students to apply for admission to each targeted U.S. university and seek work on the targeted project.

Taken together with other factors, this situation can result in an overall failure to protect IP and undermine U.S. research competitiveness. (Although a culture of openness and the presence of foreign students are generally considered strengths of the U.S. system, in this context such factors could represent a challenge to capturing the full value of U.S. investments.)

Finally, one participant said that a small or mid-sized U.S. company may find that it is competing with an overseas, state-run company—essentially meaning that a small or mid-sized U.S. company is competing against another nation.

PARTICIPANTS CONSIDERED WAYS TO ENHANCE U.S. COMPETITIVENESS

In the context of the challenges to U.S. competitiveness described in the previous section, forum participants generally agreed that there is a need for action, including government action. Specifically, participants agreed that the United States should

- increase its participation in international efforts to set basic standards for nanotechnology, as explained below, and
- act to enhance U.S. competitiveness through one or more of the following approaches—(1) strengthening innovation across the U.S. economy, (2) promoting innovation in U.S. manufacturing, and (3)

designing a grand strategy to reach a vision for U.S. nanomanufacturing.

Some actions might be taken by the U.S. government acting alone, while others would likely involve a range of actors, including the federal government as well as, for example, industry, academia, and state governments.

Participants Saw a Need for Increased U.S. Participation in Setting Basic International Standards

Several forum participants emphasized the importance of international cooperation in setting basic standards for nanomaterials and products that are enhanced by nanotechnology.[44] Governments as well as the scientific community and the private sector have an important role to play in this regard, as international standards can facilitate expanding international trade in nanotechnology-enhanced products. While the U.S. model of setting standards reserves a larger role for leadership from non-government stakeholders, U.S. government support for the scientific and business communities' efforts in this regard can help avert the risks associated with a multiplicity of individual-country standards, or with international standards that were set by other nations without sufficient U.S. input. Importantly, however, multiple forum participants emphasized that currently, there is a need to increase U.S. participation.

Accordingly, basic international standards are important for expanding the economic opportunity for emerging technologies. Lack of uniform international standards can result in limiting the economic opportunities of companies with superior technologies because a multiplicity of standards can effectively result in a multiplicity of fragmented markets. Without a basic degree of standardization, a manufacturer of a certain product has to contend with higher costs and more complex logistics in order to sell its products in markets with widely different standards.[45] One forum participant emphasized this point by saying: "Without standards, we cannot sell anything."

One participant explained that setting basic standards includes characterization of materials and naming and measurement conventions. While regulatory standards dealing with environmental, health, and safety issues related to nanomaterials and nanotechnology-enhanced products are discussed in a subsequent section of this report, we note here that forum participants generally agreed that important economic risks are associated with the inadequacy of the knowledge base about nanomaterials and products that contain them. For example, some forum participants expressed the view that consumer trust in a new product is essential for its success. Basic standards can greatly facilitate the accumulation of a knowledge base that is necessary for greater transparency in markets (and can also facilitate progress in addressing the environmental, health, and safety aspects of nanomanufacturing, as discussed in a subsequent section).

Several forum participants emphasized that U.S. researchers in the field of nanotechnology are limited in their ability to participate in international meetings to discuss standards because of lack of government funding or alternative budget priorities. Some pointed to the risk that U.S. nano-innovators might have to contend with other countries' taking the lead in developing international nanotechnology standards, leaving the U.S. behind. That is, the outcome could be the development of international standards with relatively little input from the U.S. nanotechnology community or the development of a patchwork quilt of diverse national standards. This in turn could make it harder for U.S. producers to compete in foreign markets.

Participants Discussed Three Approaches to Enhancing U.S. Competitiveness in Nanomanufacturing

Forum participants considered ways that the United States could address challenges to its competitiveness (in addition to standards issues). As described earlier in this report, key challenges to U.S. competitiveness cited by participants include:

- gaps in U.S. funding, investment, and support (that is, the *Valley of Death* and the *Missing Middle*), which may hamper transfer from R&D to large-scale nanomanufacturing;
- issues such as prior offshoring and possible U.S. workforce limitations, which may challenge our ability to establish and maintain nanomanufacturing in the United States;
- lack of a national vision for U.S. nanomanufacturing and limited technology transfer at some U.S. universities; and
- increasing global competition (for example, other countries are addressing funding or investment gaps) coupled with the fact that some countries, or entities within them, are "playing by new rules."

Table 2. Three Approaches to Enhancing U.S. Competitiveness, with Proposed Actions and Rationales

Approach	Proposed actions	Rationale
1. Strengthen innovation across the U.S. economy	Continue or update federal policies and programs that help strengthen innovation generally (i.e., across all sectors of the economy).	Government acts to supply goods and services critical to innovation when private markets fail to do so, most often because firms cannot capture the full benefits of providing them. Beyond these measures, which include providing education and building infrastructure, firms are in a better position than government to make decisions about how to allocate resources to the most promising innovations.
2. Promote innovation in U.S. manufacturing	Establish centers, encourage clusters, or design programs to address the *Valley of Death* or the *Missing Middle*, especially as applied to manufacturing, with financial support coming from public-private partnerships. (We note that this approach may be applied to specific areas within manufacturing, for example, nano-manufacturing, as indicated by the examples given in the below discussion of this approach.)	The United States needs a strong manufacturing base because it is essential to the economy and to innovation itself. Moreover, structures separating manufacturing from design can have significant adverse results. Assuring this base means "leveling the playing field" in the global economy—by directly addressing the *Valley of Death* and the *Missing Middle*, especially as these apply to innovative manufactured products or innovative manufacturing processes.
3. Design a "grand strategy" for U.S. nano-manufacturing	Define a vision for U.S. nanomanufacturing. Design a grand strategy (based on a systems approach) for achieving this vision—through a collaborative process that might be led by the federal government.	Nanomanufacturing is a megatrend: Nano-manufacturing will significantly affect future U.S. competitiveness in global markets as well as providing societal benefits. Nanomanufacturing may be a future general purpose

Approach	Proposed actions	Rationale
		technology (GPT) akin to digital technology or electricity, and thus it could be classified as a public good with anticipated benefits for the entire economy—potentially justifying targeted federal support. Moreover, nanomanufacturing may become an engine of job creation as its disruptive innovations proliferate throughout the U.S. economy.

Source: GAO analysis of forum information.

Taken together, these challenges can represent an emerging threat to the United States' ability to realize a level of future economic benefits commensurate with its investments. Participants therefore considered approaches to enhancing U.S. competitiveness in this area. Three major approaches—which might be viewed as either representing alternatives or complementing each other—emerged from the discussion and are briefly outlined in table 2. For each approach, participants discussed proposed actions and a supporting rationale.

Participants did not voice objections to strengthening innovation across the economy (Approach 1), although some did not view this approach as sufficient by itself. That is, some urged more targeted action, as represented by Approaches 2 and 3. Therefore, each approach is described in more detail below.

Approach 1: Strengthen Innovation across the U.S. Economy

Federal policies and programs that strengthen innovation across the economy represent an established U.S. approach. Proposed actions and a rationale for this approach are discussed below.

Proposed Actions: Continue or Update Efforts to Strengthen Innovation across the Economy

This approach does not focus on supporting any one type of technology or any particular economic sector. Rather, government action is typically motivated in the presence of spillovers, when an activity generates benefits that a private firm cannot fully capture. In such circumstances, firms will not supply as much of that activity as would be desired from society's point of view. Such activities include the following:

- education, which can help assure a future workforce that will meet needs as generations change;[46]
- development of human capital, that is, the knowledge and skills of the workforce, for which continuing education may be crucial because technological change often brings changes in job skills requirements;[47]
- immigration policies, which may provide the country with both skilled workers and potential entrepreneurs;
- frameworks provided by economic policies, including incentives for innovation and the removal of disincentives to innovation, management of growth and stability in the macro economy, promotion of open trade and competition, protection of intellectual property rights, nurturing of vibrant capital markets, administration of an efficient tax regime, and vigilance to assure effective regulation;
- research with anticipated spillover benefits, including basic research, the development of measurement techniques and databases needed for research, and so forth; and
- infrastructure support, which has been a key part of U.S. strategies to promote economic growth and includes not only the transportation system but also, more recently, knowledge-oriented infrastructures, such as broadband telecommunications.

Although such actions are long established, the specific activities or efforts to enhance competitiveness in innovation may change over time. Potentially, existing efforts could be updated or improved in light of current and anticipated opportunities and challenges.

One participant objected to limiting U.S. action to strengthening innovation across the economy. Specifically, he said that the historical development of the U.S. economy has included targeting support to specific industries—and mentioned radios, aircraft frames and engines, semiconductors, and computers. His view was that targeted support in instances such as these had helped the foundations of the entire economy. He also noted that other countries have targeted support to specific industries and succeeded. However, an explicit rationale for limiting support to the kinds of actions described above was presented and is reprised below.

Rationale for Strengthening Innovation across the Economy

The established policy approach described above (supporting innovation across the economy) has underpinned innovation and economic growth at least since the end of World War II. Support for individual sectors has most often arisen out of the government's own needs (e.g., aerospace and defense). The admonition that the government should not "pick winners and losers" stems from the likelihood that the government will not have sufficient information or foresight about an individual firm's or a particular technology's growth potential to select it for special subsidy.

This view emphasizes that the U.S. political and economic system is based on private-sector primacy and advocates allocating resources through the functioning of market mechanisms—because such mechanisms are anticipated to result in U.S. investments that not only are the (1) most efficient, but also (2) given today's global content, best suited to the comparative advantages of the United States vis-à-vis other nations.

To illustrate how the market-centered viewpoint applies to innovation, one forum participant said that early government funding of a particular technology risks giving it an unfair advantage over others that might prove to be more technically and commercially feasible. In other words, in his view, greater competition among different innovations has a better chance of yielding innovations that are best suited for manufacturing scale-up and acceptance by markets. Briefly stated: according to advocates of this approach, the federal government ought not select or single out certain firms, sectors, or technologies for support; in fact, it should not choose manufacturing over, for example, services.[48]

The current perceived lack of market support for investment in nanotechnology (mentioned in an earlier section of this report) led to a discussion about whether increased market support would be forthcoming in the near future. Some participants expected that such support would be forthcoming if there were either (1) a near-term ending of the current VC focus on Internet services (which some hypothesized is a "bubble"), followed by a "tectonic shift" of VC toward new materials and nanomanufacturing, or (2) a widespread realization by major U.S. companies and corporations that nano-enhanced products will make their current products obsolete—resulting in decisions to invest substantial amounts in nanomanufacturing commercialization efforts relevant to their businesses.

Approach 2: Promote Innovation in U.S. Manufacturing

Forum participants expressed concern about the *Valley of Death* and the *Missing Middle*, as discussed in an earlier section of this report. They also noted the "friction" involved in commercializing innovative nanotechnology products or processes.[49] Specifically, they felt that finding ways to reduce friction for U.S. innovators would both shorten time frames for innovation and lower associated costs—and as a result, lessen risks to the commercialization process. Proposed actions and a rationale are discussed below.

Proposed Actions: Establish Centers or Programs to Address the Valley of Death and the Missing Middle

Forum participants identified two centers and a pilot program aimed at developing ecosystems or infrastructures to create the conditions for innovators to more successfully traverse the *Valley of Death* and the *Missing Middle*, as described below. These examples alternatively focus on (1) nanomanufacturing for several applications; (2) nanomanufacturing as it applies to the semiconductor industry; and (3) advanced manufacturing in general.

Example 1: NASCENT at the University of Texas at Austin. The Center for Nanomanufacturing Systems for Mobile Computing and Mobile Energy Technologies (NASCENT) was founded at the University of Texas at Austin in September 2012, with funding from NSF. Its objectives are to (1) create processes and tools for manufacturing nano-enabled components for mobile-computing, energy, healthcare, and security—as well as simulations for testing potential nanomanufacturing approaches—and (2) provide an ecosystem with computational and manufacturing facilities (for example, large-area wafer scale and roll-to-roll nanomanufacturing[50]), as well as the university's resources, including faculty, staff, and students. The overall goal is to facilitate the rapid creation and deployment of new products and to mitigate the risks associated with the *Valley of Death* and the *Missing Middle*. Another goal of the center is to use "10 years of NSF funding to develop the center infrastructure so it will ... [become] self-supported from industrial partnerships and other [non-NSF] funding sources." Center partners include

- industrial partners—such as tool makers, materials suppliers, and device makers—who will provide both technical and financial support;
- companies ranging from start-ups to well-established firms that will implement or adopt technology created by the center; and

- "translational research partners" such as technology incubators and technology funds.

Example 2: The College of Nanoscale Science and Engineering. The College of Nanoscale Science and Engineering (CNSE), established in 2004, is part of the State University of New York and is located in Albany—within the existing regional (Hudson Valley) ecosystem centered on the semiconductor industry. CNSE is designed as a unique research, development, prototyping, and educational public-private partnership for advancing nanotechnology. A chief CNSE partner is SEMATECH—a global consortium of major computer chip manufacturers that coordinates cutting-edge R&D projects on semiconductors and is headquartered at CNSE. CNSE has more than 300 members and strategic partners that include large U.S.- and non-U.S.-headquartered private companies such as IBM, Intel, Samsung, and Global Foundries; small and medium companies; universities from across the United States; and regional community colleges and economic development organizations—as well as government-agency sponsors. CNSE facilities allow the development of semiconductors just short of mass production; this is relevant for companies attempting to transition from an innovative concept to a prototype and to prepare for large-scale production. CNSE has developed models of pre-competitive collaboration among its partners, who use high-tech CNSE equipment that would be too costly for many individual companies to purchase.[51] To illustrate potential mutual benefits to participating companies, a forum participant said that one company working on developing a new product or process using CNSE facilities may be unaware of a proprietary process developed by another CNSE partner—which could help it overcome an engineering hurdle; in this case, CNSE staff might suggest using the proprietary process under a leasing arrangement from the owner. According to a forum participant, New York State's share of the investment was initially much larger than that of its private partners, but over time private funding has come to exceed New York State's funding by a factor of 16 to 1.[52]

Example 3: The Proposed National Network for Manufacturing Innovation. The proposed National Network for Manufacturing Innovation (NNMI) is a multi-agency federal pilot program, announced in 2012, that uses public- private partnerships to enhance manufacturing innovation in the United States. Specifically, NNMI aims to:

- address challenges associated with the transition from research and the early stages of development to full-scale manufacturing—by providing shared-use facilities that support regional needs for scaling up the production of goods including manufacturing research and workforce development,[53] and

- encourage the location of advanced manufacturing in the United States—thus strengthening the nation's manufacturing, design, and innovation capacity and countering further offshoring of skilled manufacturing jobs.[54]

NNMI calls for regional groups—with partners including private-sector organizations, universities, community colleges, state economic development offices, not-for-profits, local governments, and other organizations—to submit proposals for establishing an institute within a specific region or geographic location.[55] The institutes are intended to help assure a nexus of manufacturing, design, and innovation (as subsequently discussed). The initial costs of establishing each institute are to be shared by the proposing group and the federal government, and the federal portion of the funding is expected to diminish over time, as private sector and/or local government partners assume more responsibility—eventually, full responsibility as the institutes grow in their ability to deliver value.

The three models for clusters or centers described above all (1) encourage the development of regional clusters or ecosystems for science-based manufacturing in the United States;[56] (2) consist of public- private partnerships focused on pre-competitive R&D cooperation among companies in related fields; (3) intend government funding to be used as seed money that will be provided for a limited number of years or diminish over time; and (4) aim to help private developers overcome what forum participants described as a problem of inadequate support for ventures in physical science or engineering (which VC might perceive as high risk), especially at the commercialization stage.

One participant said that the effectiveness of such initiatives has not been firmly established; that successful, spontaneous clusters arise out of mechanisms that are not well understood; and that building on those that already exist might be a more conservative and possibly more effective strategy. This participant pointed to a report from the Organisation for Economic Co-operation and Development (OECD 2007), which discusses improving cluster-promoting policies and that cautions against inefficient government intervention. However, varied participants voiced a specific rationale for the approach described above, as follows.

Rationale for Supporting Innovation in Manufacturing

Views supporting the establishment of clusters, centers, and programs that address the *Valley of Death* and the *Missing Middle* include three key arguments:

1. Through history, the United States has supported development of innovative technologies, spawning new industries in the process.
2. Today, there is a need to "level the playing field" for U.S innovation-oriented businesses (especially, manufacturers), which are put at a disadvantage by other nations' successful efforts to grow their own technology-based industries by supporting innovation all the way through manufacturing scale-up and commercialization.
3. A strong U.S. manufacturing base is needed to support innovation, as described further below.

Varied participants said that:

- As mentioned previously, the United States has a long history of providing significant support for specific innovative industries, including the radio, the semiconductor industry, the computer, and aircraft technologies. According to this view, the U.S. economy reached its overall highly competitive global position, in part, *by targeting support to specific industries* in at least some instances. (Often, such support has been related to U.S. defense needs.)
- Today, other nations' governments are supporting innovative businesses in areas such as nanomanufacturing or addressing specific challenges to innovation—such as the inadequacy of funding or other resources that occurs at points of transition from R&D to actual products or large-scale production. To cite two examples, participants mentioned (1) Germany's Fraunhofer-Gesellschaft— Europe's largest research organization—which is viewed as a successful public-private partnership that helps incorporate technological innovation into different industrial sectors; and (2) Russia's government-owned company, Rusnano, which manages a multi-billion-dollar fund focused on acquiring nanotechnology-based businesses from around the world. (According to some observers, Rusnano has acquired some companies that failed to receive the support that they needed in their own countries, for example, the United States, where the underlying technologies were first developed.) Additional challenges from varied

countries around the world are discussed in an earlier section of this report.

- There is a need, from this global perspective, to establish and maintain a "nexus" of manufacturing, design, and innovation (MDI), which is tied to the need for a strong U.S. manufacturing base. A National Academy of Engineering report described the intersection of manufacturing, design, and innovation as delivering value that is enabled by a physical product (Whitefoot and Olson 2012). This concept posits that in many, perhaps most industries, it is not possible to effectively disaggregate the manufacturing process from the design of new materials, new processes, and new products. Structures that separate manufacturing from design (as occurs when U.S. companies offshore most or all their manufacturing but attempt to retain design here in the United States) can have a significantly adverse effect.[57] This adverse effect reflects the earlier mentioned statement by a forum participant who, speaking of advanced batteries, said: "When we design here [and] we ship [manufacturing] abroad, we lose this shop-floor-innovation kind of mentality." In sum, the MDI view holds that offshoring of manufacturing jobs can put the United States at a disadvantage in terms of its ability to (1) maintain design expertise and (2) achieve innovation. From this perspective, a strong manufacturing base is what supports innovation.

Further, some participants saw a strong manufacturing base as expanding opportunities for quality employment in the economy. A case in point—according to one participant—is the role of CNSE in Global Foundries' decision to locate a major manufacturing fabrication facility in the Albany region. Global Foundries, a leading global manufacturer of semiconductors, established a manufacturing facility near Albany, New York, in order to benefit from the R&D environment and the supply chain and infrastructure advantages that the region offers. According to this forum participant, the Global Foundries semiconductor fabrication facility brought over 1,000 high-quality jobs to the region.

Additionally, the rationale for this approach (innovation in manufacturing) holds that a strong manufacturing base may generate additional activity in the services sector. An example of this was provided by a forum participant who explained that a large share of IBM's revenues derive from services, but these services are based on producing mainframes and other hardware that, in turn, are based on proprietary microelectronic devices. Figure 12, below, illustrates IBM's "inverted stack business model."

Approach 3: Design a U.S. Grand Strategy for Nanomanufacturing

This approach focuses on a vision or goal for nanomanufacturing innovation and uses "systems thinking" to devise a comprehensive strategy designed specifically to achieve that vision or goal. Systems thinking is a holistic approach to evaluation, analysis, and strategy that considers how the interdependent parts of a system work together—and within a larger context, over time.[58] As a forum participant said, the U.S. federal government laid the foundation for a systems approach when it established the National Nanotechnology Initiative (NNI) a decade ago.

Proposed Actions: Define a Vision for Nanomanufacturing and a Grand Strategy to Attain It

As outlined by a forum participant, the full strategy would be designed, proceeding from a vision or goal to the examination of the social, technological, economic, environmental, and political elements of the relevant systems and their interactions with one another; understanding the basic science, engineering, and manufacturing involved; and consulting the full range of stakeholders. This participant said that although systems thinking and the design of a grand strategy, based on a vision, are often employed following a crisis that motivates a nation, this approach can be usefully pursued in advance of a crisis, using foresight. This approach reflects the statements of one participant who said, in effect, that the future of nanomanufacturing for the United States is limited only by our ability to envision what we want to see realized. This approach would likely draw upon the U.S. federal government to develop and articulate the strategy—in coordination with industry, academia, nonprofits, and state and local governments. Additionally, some federal effort is implied for implementation, but the level of funding and the mix of funding sources (not specifically discussed at the forum) would likely be specified as part of developing a vision and strategy for nanomanufacturing.

Participants discussed potential elements of this grand strategy, which could encompass a diversity of actions by federal or state governments, as well as academia and industry, such as: (1) addressing the *Valley of Death* and the *Missing Middle*; (2) possibly changing the tax structure— for example, to incentivize private-sector long-term investments (e.g., 5 to 10 years or longer) by lowering tax rates, eventually to zero tax;[59] (3) keeping up to date on what other countries are doing as input to strategic considerations, going forward (not necessarily to emulate other countries);[60] (4) partnering with the European Union or key nations; (5) articulating substantive national goals that nanotechnology might help achieve and holding "grand challenge" contests with large cash awards;[61] (6) encouraging the professionalization of technology transfer in universities across the country (potentially using best practices of technology transfer offices in leading universities as examples); (7) breaking down silos in universities to encourage innovation based on interdisciplinary work and goal-focused collaboration; and (8) possibly expanding selected programs (such as NSF's Innovation Corps) that encourage basic-science or basic-engineering investigators to "extend their focus beyond the laboratory" and translate findings into commercial results or benefits to society.[62]

Still other actions were suggested at the forum or in post-forum comments, including:

- convening industry-led consortiums to conduct road mapping and tool building to reduce barriers in expanding U.S. nanomanufacturing, and
- addressing nanomanufacturing as part of a larger international dialogue (for example, the Transatlantic Trade and Investment Partnership or TTIP that is now being considered[63]).

One participant cautioned against pursuing a grand strategy based on anticipated job creation. This participant noted that in general, advanced manufacturing does not appear to be labor intensive and that salaries for manufacturing jobs have not increased in line with productivity increases—observations which imply that even with new markets, subsidies for nanomanufacturing may not result in large numbers of well- paying jobs. However, a specific rational for developing a grand strategy for nanomanufacturing was discussed, as described below.

Rationale for Developing a Grand Strategy for Nanomanufacturing

Multiple participants expressed varied views that, taken together, argue for federal support of nanomanufacturing as a special case, attributable to the belief that such technology has the potential to generate benefits beyond those that accrue to firms in that industry.

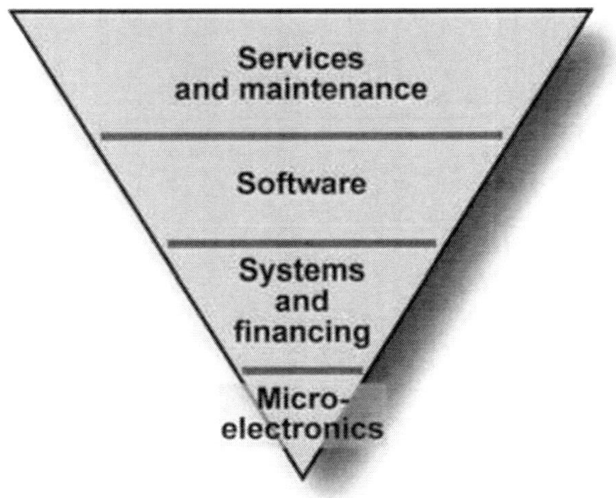

Source: GAO adaption of graphic provided by IBM.

Figure 12. IBM's Business Inverted Stack Business Model.

Earlier in this report, forum participants were described as generally viewing nanomanufacturing as a potential future megatrend that could bring societal and economic benefits equal to or beyond those provided by digital technology—and that could potentially have major impacts on the future competitiveness of the United States. Multiple participants also noted that nanomanufacturing enhances products across many economic sectors. Thus, as one argued, federal support for nanomanufacturing should not be viewed as supporting a single industry; that is, it spans many industries. In this sense, some viewed nanomanufacturing as, in effect, a possible future "general purpose technology" (GPT) that, as mentioned earlier, might eventually provide benefits as widespread as those associated with the introduction of electricity, thus possibly qualifying as a public good. For example, Thomas Edison and Nikola Tesla, who played pivotal roles in advancing 20th century technology based on electricity, could not possibly have acted to produce the scope of societal benefits ushered in by their inventions; rather, public investments in the electric grid allowed the enormous benefits of electricity to spread to society as a whole.

Multiple participants also identified nanomanufacturing as a future disruptive technology or innovation that could open new markets and result in net jobs creation—at least for successfully competing nations. From this perspective, nanotechnology innovations would not just raise productivity for firms in existing markets but also create new markets. Finally, manufacturing jobs in general have been reported to be higher paying than nonmanufacturing jobs and more likely to provide workers with benefits.[64]

Participants Encouraged an Integrated Framework to Assess and Address EHS Implications of Nanotechnology

Forum participants offered a wide range of perspectives on the environmental, health, and safety (EHS) implications of nanotechnology, nanomanufacturing, and nanomaterials. Forum participants presented information on what is currently known about these implications and expressed frustration about the lack of progress in understanding the risks from potential exposure to nanomaterials. Participants specifically noted a current dilemma related to identifying or determining EHS risks. Because so few nanomaterials have been studied and no long-term or chronic data are available, it is very difficult to predict and manage risks for new nanomaterials. Forum participants also identified significant research needs to discern EHS implications, and they discussed the need to fully communicate the benefits and risks of nanotechnology to the public, helping to distinguish between perceived and real risks. While participants noted an underlying tension between advancing innovation in nanotechnology and adopting regulation to address any negative EHS implications, they suggested that an integrated EHS framework and a collaborative approach could help offset these tensions. Forum participants discussed developing an EHS framework that would help build safety into product design and include industry, academia, nonprofits, and government.

Participants Expressed Frustration about Lack of Progress in Understanding EHS Implications of Nanotechnology, Nanomanufacturing, and Nanomaterials

Some forum participants expressed frustration that EHS implications have been discussed for roughly 10 years and yet little is known about the risks from potential exposure to nanomaterials. Some participants noted that only a few types of nanomaterials have been studied and that the universe of nanomaterials available in the market is unknown given the range of nanomanufacturing activities. However, a participant presented research that concluded that a number of characteristics or properties of nanomaterials may contribute to EHS hazards and that workers, in particular, may be exposed to nanomaterials.

One presenter noted that while hazardous effects have been seen in certain nanomaterials, only certain types of nanomaterials have been studied and that we lack the ability to predict hazards for nanomaterials generally. This presenter provided exposure data that showed risks are higher for workers than the general population, since material used in production is in a freer form, but presenters also discussed how clean rooms and other practices could reduce workplace exposures.[65] One presenter also stressed that most EHS research to date has been animal- based and that very little data for human exposure have been collected. Another presenter stated that there is a fundamental lack of data and information for identifying and quantifying the EHS impacts through the life cycle of the nanomaterial, and that the U.S. government alone cannot generate this data.

Participants debated the risks that nanomaterials pose, with some suggesting that risks could be categorized as perceived rather than real. A few industry participants explained that the nanomaterials used in their products are inert and pose no potential risk of exposure. One of these participants emphasized that nanomanufacturers have been using known approaches for protecting workers' health, such as technologies used in clean rooms and those developed for handling dangerous chemicals. A few participants emphasized the concept of responsible development of nanotechnology, with one saying it is an essential component for nanomanufacturing. Such an approach requires acknowledging possible hazards and taking precautions to prevent exposure to them until more detailed information is developed.

Significant Research Needed to Discern EHS Implications

Participants discussed some similarities and differences between nanotechnology and prior technologies. One participant noted that nanomaterials have been used for a long time, although our understanding and control of them is now much more sophisticated. However, participants also discussed the wide range of new materials and products that nanotechnology could create. Participants described nanotechnology as generating new questions about EHS impacts.

Participants said significant research was needed to discern or anticipate EHS implications of manufacturing with nanomaterials and using nanotechnologies. Participants noted the presence of significant funding—both governmental and private—for nanotechnology research, but one participant suggested relatively little funding supports research on EHS implications, an observation that is consistent with our previous reporting on the National Nanotechnology Initiative.[66] Some forum presenters pointed out significant challenges for EHS research stemming from the very complex science involved, including the innumerable chemistries of various nanomaterials. For example, one participant explained that the universe of materials is diverse and heterogeneous. In addition, exposures in the workplace have not been uniformly measured or widely characterized, and measurements to date have been related to specific tasks and locations. Another participant noted that the effects at the quantum level are less predictable and therefore less defined. Some participants noted the significant information gaps that EHS research could help fill, for example, to determine where the highest risks of exposure existed. While acknowledging research challenges, one presenter made the comparison with weather simulations for hurricanes, where improving the models used in weather prediction has had a dramatic impact, and suggested that computational tools are necessary for modeling complex nanomaterial risks. In addition to increased computational capacity, multiple participants discussed sustainable design and "green chemistry" as potential areas upon which nanoscience research could draw or from which it could benefit in considering EHS implications. One participant suggested that an increased focus on responsible development of nanomaterials could alleviate international trade disputes over precautionary regulation.[67]

The identification of research needs and assessment of progress toward meeting those needs is also the subject of a National Research Council (2013) study. This study identified needs in areas such as modeling nanomaterial releases along the value chain and developing instrumentation to measure key nanomaterial properties and changes in them in complex biologic and environmental media.

Participants also discussed differences between considering the EHS implications of therapeutic uses of nanomaterials and the occupational safety and health implications of nanomaterials. When designing therapeutic uses, one participant noted the significant resources expended on investigating nanomaterials and performing clinical trials. However, worker health and safety issues may be more challenging since workers are the first people in society exposed to the products of any technology. A few participants stated that a company whose priority was to bring a new product to market as quickly as possible and profitably, may rush to market and risk EHS implications. Thus, the underlying research for worker safety may not be an individual company priority. However, one participant said that many protective measures already used to reduce workplace exposures to larger, aerosolized particles— such as controls often used in the pharmaceutical industry—will also apply to nanoparticles.

Multiple participants spoke about the need for long-term commitment for EHS research. One participant stated that EHS research would need to be carefully orchestrated and thought through; that it would not be a 2- year project. For example, multiple participants stated research would need to consider new approaches to the assessment of product life cycles—that is, to go beyond just disposal, but also include recycling of materials.

Participants Recognized Tension between the Rapid Introduction of New Products and the Need to Discern EHS Implications

While multiple participants conveyed that companies could mitigate potential hazards to gain benefits from product innovations, some feared regulation to address EHS concerns could damage U.S. competitiveness and others noted the need for precaution when dealing with nanomaterials. For example, one participant noted that a company had designed an easier-to-clean scope for colonoscopies using nanotechnology. While the company assumed it would be an easy approval, the regulatory agency asked it to run a clinical trial to prove the new scope was safe and effective. Estimates for clinical trials of similar medical devices have been projected to be $30 million, which the participant suggested raised questions about whether investors could be found. One participant also raised the concern that constituencies (in other countries) that have been developed with the precautionary principle would significantly delay the continued expansion of nanotechnology, that is, until the risks are known and mitigated. Another participant, noting global competitiveness issues, worried that early but lengthy standard-setting efforts might put U.S. industry on hold at a very competitive and time-sensitive stage of the innovation process—while other nations progressed. In contrast, one participant stated that there is enough evidence of health effects for workers that prudent precautionary practices are warranted and effective, and could be a positive factor in business growth. For example, one participant noted that his company's product was manufactured in a clean room environment and included a nanomaterial bonded into a larger aggregate. Some participants also discussed how companies perform research on the safety of their nanomaterials and their efforts to inform regulators on the science behind their products. Others noted that innovation and regulation in relation to nanomaterials and technology would have to be considered as part of an evolving environment that extends beyond traditional research and industry settings. Some forum participants pointed out that technology involving nanomaterials has been evolving and that new implications for EHS could arise. One participant pointed out the new desktop technologies and the democratization of manufacturing raised concerns that certain nanomaterials (for example, those in powder form) could be inhaled and lead to widespread exposure without users' knowing about and managing the potential hazards. For example, consumer use of desktop 3-D printers could rapidly and widely advance innovations, but controlling the associated EHS risks for individual homes or garages would present major regulatory challenges.[68]

Participants Saw a Need to Raise Public Awareness and Understanding of Benefits and Risks

A few forum participants presented information that suggested that the public does not understand nanotechnology and likely perceives risks from nanomaterials. For example, one participant noted that based on some public surveys and statistically valid samples, his or her research found that perhaps 40 percent of people had heard the term "nano." Because of this lack of awareness, the public's perception is going to be dependent on the application and how the benefits are presented. This participant noted that the public does not have a sense of differentiation about nanomaterials—some may be considered high value, for example, materials for a better colonoscopy, but others address everyday concerns, such as stain-resistant ties and pants.

The forum's participants discussed the need to educate the general public on nanotechnology and that companies should fully consider how they develop communication strategies about the potential hazards associated with using such products. For example, one presenter's research pointed out the public's basic distrust of where to seek information, including from both industry and government. Some participants noted that this concern will need to be considered when determining how to educate the public on the benefits and potential risks from new products developed from nanotechnologies and nanomaterials. One participant also noted that the legal community would have to develop scientific literacy to implement and communicate scientifically sound legal standards. One participant from the business community wondered whether to continue referring to nanotechnology in product names, advertising, or labels—given that "nano" may convey risks that actually might not apply. Participants also discussed public reaction to genetically modified organisms as highlighting the need to understand and communicate about nanotechnology risks.

Participants Encouraged a Collaborative Approach for Developing an EHS Framework

Multiple forum participants emphasized during their discussions that a collaborative approach—to include industry, academia, nonprofits, and government—could help realize the benefits of nanotechnologies and associated products and mitigate EHS risks. Participants suggested ideas for and roles within such an approach. For example, industry could identify which products use nanotechnologies or incorporate nanomaterials and disclose what it had learned concerning those materials. Academics could help catalogue such information in a way that would be useful for designing new products or create frameworks that companies could consider during product design and when conducting life-cycle assessments. In addition to supporting education and research programs, government could work with industry and academia to set appropriate standards and support public-private partnerships and international collaborations. One participant stated that government involvement in collaborations between academics and industry is instrumental to their long-term success.

A few of the participants wanted a stronger role for industry to assist in the identification of nanomaterials and their uses, as well as disclosure of any EHS implications learned. Specifically, participants pointed out that it is challenging to identify the range of uses of nanomaterials in consumer and commercial applications without the input of the companies involved in the development and production of these materials. Since there are few tools and methods to predict the behavior of nanomaterials in the environment, industry is in the best position to help identify which materials are in use. A few participants noted that industry could explain the research behind the product and its real-world impact so that others could understand and distinguish real from perceived risks.

A few participants emphasized a stronger role for government in setting standards for nanotechnology and nanomanufacturing. While some industry participants feared potential regulation would stifle competitiveness, forum participants discussed the need for some standardization, especially in defining basic terminology, in order to characterize specific materials and products across global markets. Some participants discussed how a standard descriptive system would enable efforts to characterize the physical attributes of materials at the nanoscale and allow greater collaboration among industry, academic, and government researchers. One participant noted that in the field of nanomedicine, a central characterization laboratory has spurred innovation and commercialization by providing standardized information on nanomedicines in an open database. A few participants suggested that this approach could be applied to the broader topic of nanotechnology products.

Multiple participants suggested developing an integrated EHS framework for thinking about nanotechnology, nanomanufacturing, and nanomaterials. One participant explained that the framework would be based on incorporating assessments of EHS implications into the design phase of the product—not at the end of life, not at disposal, and not after problems or health impacts to consumers or workers have already occurred. Participants characterized this concept as "safer by design." One participant explained the idea as capturing the functionality of the product while addressing safety concerns. Participants also discussed the importance of considering the life cycle and conducting life-cycle material assessments. Such an assessment would consider not only the use of the material, but all stages of the product's life cycle from production and development through disposal and recycling. The following illustrates an example of a life-cycle assessment.

One participant expressed the opinion that the United States lacks a coherent governance and oversight system for nanomaterials and nanotechnology, a lack that could be problematic for U.S. industry and innovation as individual municipalities or other countries put governance structures in place. One participant suggested that the nations that complete standards and risk management systems first will have an advantage in supporting development of new nanotechnology products and companies. Others specifically cited the European Union's precautionary approach and required labeling as at least reducing uncertainty in how such products are regulated in that market. One participant noted ongoing participation in international cooperative efforts over the last 8 years in about 10 different global standardization efforts with respect to nanotechnology. Another participant discussed one global effort at the OECD to develop approaches for responsible development of nanoscience and nanotechnology.[69] We previously reported (GAO 2013b) that early and ongoing coordination with foreign governments in emerging areas before regulations are in place may facilitate international regulatory cooperation. While some participants considered the National Nanotechnology Initiative a productive government effort, participants also noted that the initiative does not have a centralized source of funding or decision-making authority.

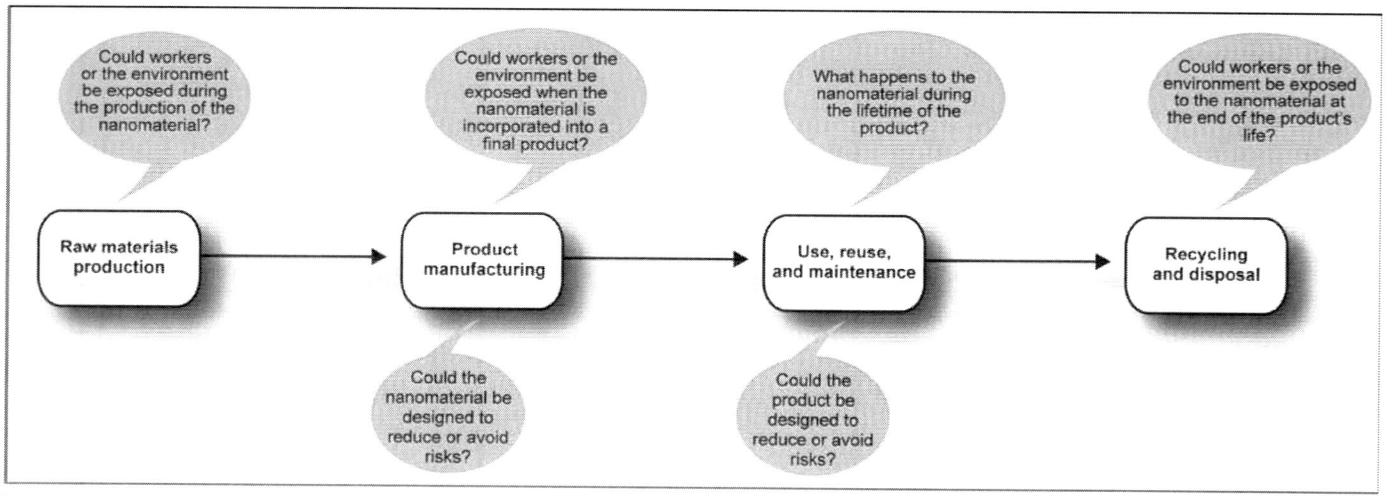

Source: GAO.

Figure 13. Illustration of Product Life Cycle and Issues Posed.

CONSIDERATIONS GOING FORWARD

Forum participants described nanomanufacturing as an emerging global megatrend: a technological revolution that will likely bring future world-changing developments, including:

- new applications across many industries and related social impacts that may match or exceed levels of change associated with the digital revolution or the advent and spread of electrical power (a view that is in line with forward-looking literature indicating that nanotechnology has the potential to qualify as a general purpose technology in the future), and
- new global-scale economic opportunities and an increasingly intense international competition.

According to forum participants, the anticipated importance of future nanomanufacturing developments suggests that going forward, the United States should consider both retaining effective existing policies (with updates as needed) and taking steps toward new actions or strategic responses that could address key gaps and challenges.

Participants Said Maintaining a High Level of Investment in Fundamental Research Is Essential

Forum participants said that it is essential for the United States to maintain a high level of investment in fundamental nanotechnology research. Two reasons are: (1) while the United States is currently viewed as the likely overall leader in nanotechnology R&D, certain other countries are now making significant investments in R&D as well as, in at least one case, publishing large numbers of papers, and (2) ongoing research breakthroughs will continue to drive the future of nanomanufacturing. Further, forum participants explained that nano- innovators may need to both

- develop the new technology or product itself—a process that typically begins with fundamental research ("early *technology* readiness levels"),[70] and

- devise a new and potentially innovative manufacturing method to mass-produce that product—a process that may begin with basic engineering research ("early *manufacturing* readiness levels").[71]

Although these two research processes—one involving technology development and the other, manufacturing development--may often be intertwined, one participant emphasized that it may be important to consider, as nanotechnology increasingly moves into manufacturing, not only (1) continuing support for fundamental nanotechnology research but also (2) targeting at least some of the funding for nanotechnology research to early-stage research on nanomanufacturing (that is, research aimed at conceptualizing innovative processes for eventually testing and mass-producing new nanomaterials and nano-enabled products and developing these processes in a laboratory environment).

Participants Identified Four Key Areas Needing Action

Forum participants identified gaps and challenges in four key areas where they believe future action is needed:

1. *International data on R&D investment:* Public-sector investments by nations have been considerable. The U.S. National Nanotechnology Initiative (NNI), begun in 2001 and focusing primarily on R&D, represents a cumulative investment of almost $20 billion including the federal request for fiscal year 2014. With respect to recent and current annual levels of R&D investment, NNI's fiscal year 2013 budget was over $1.5 billion, and some other nations are now making public-sector investments that may surpass that figure. Private-sector R&D investments are also significant in some countries, including the United States. Overall, the United States currently appears to be the lead-investor nation. However, two key participants cited data reliability issues. According to one of these, a pathway forward might include actions such as convening international conferences on tracking public-investment data and other related data (such as program evaluation data), with representatives from key governments from around the globe.
2. *International standards:* Forum participants said there is a lack of basic, agreed-upon standards to facilitate industry progress in

nanotechnology, international trade, and potentially, appropriate labeling of nano-enabled consumer products. Progress on basic standards may also help address other challenges and gaps discussed in this report. Importantly, forum participants said there is currently insufficient effort, especially by the United States, to participate in and "jump start" standards development. Notably, participants said that in the currently restricted U.S. budget environment, federal agencies have appeared not to prioritize staff travel to participate in international conferences—and that it is important to remedy this situation with respect to nanotechnology standards.

3. *U.S. competitiveness:* Participants recognized that current challenges to U.S. competitiveness in nanomanufacturing across world markets, taken together, represent a serious threat to realizing a level of future economic benefits commensurate with U.S. investments. These challenges range from U.S. gaps in funding or investment for nano-commercialization (the *Valley of Death* and the *Missing Middle*) and issues such as prior offshoring and possible workforce education and training issues—to the lack of a U.S. vision for a nanomanufacturing capability. Also relevant are issues concerning knowledge about and recognition of practices of other countries that may be key to global competition and may, in some cases, constitute threats. Participants outlined three possible approaches to enhancing U.S. competitiveness:
   - updating federal policies aimed at supporting innovation across the economy (for example, investments in infrastructure and education), which is a long-standing approach;
   - encouraging or facilitating public-private partnerships that specifically address the *Valley of Death* and the *Missing Middle* in advanced manufacturing and innovation, a step that could help support a strong manufacturing base in the United States (although the examples provided by participants include centers that focus specifically on nanotechnology or nanomanufacturing); and
   - defining a national vision and designing an overall grand strategy for U.S. nanomanufacturing—an approach that might be justified if nanomanufacturing is deemed a potential or likely future general purpose technology.

   These three approaches to enhancing U.S. competitiveness might be considered alternatives, or two—or possibly all three—approaches might be used together.

4. *EHS issues:* Participants indicated that currently limited research, including a lack of data on the long-term or chronic EHS impacts of new nanomaterials, makes it difficult to predict and manage relevant risks—and difficult to help the public distinguish between real and perceived risks. The underlying tension between advancing innovation in nanotechnology and adopting regulation to address any negative EHS implications represents another possible difficulty. Participants also indicated that to advance in this area would require a revitalized approach that is integrative and collaborative.

While participants noted that action in each of the four areas above deserves consideration, these areas—and future efforts to address them, if made—may overlap. For example, basic international standards that set definitions to facilitate industry progress and trade could also help advance efforts to achieve more comparable international-investment data. Such overlap could serve as the basis for the development of a coordinated framework for nanomanufacturing-related issues.

APPENDIX I: FORUM AGENDA

Comptroller General Forum on Nanomanufacturing

Hosted by the U.S. Government Accountability Office with the Assistance of the U.S. National Academies

	DAY ONE: TUESDAY, JULY 23, 2013
12:45 p.m.	ARRIVALS/CHECK-IN
	OPENING SESSION
1:00 p.m.	**Welcome** The Honorable Gene L. Dodaro, Comptroller General of the United States, U.S. Government Accountability Office
1:10 p.m.	**Overview** Dr. Timothy M. Persons, Chief Scientist, U.S. Government Accountability Office
1: 35 p.m.	**Significance of Nanomanufacturing to Congress and the Nation** The Honorable George Allen, former United States Senator and former Governor of Virginia (10 min.) The Honorable Bart Gordon, former Member of the United States Congress; Partner at K&L Gates (10 min.)

	U.S. INVESTMENTS AND COMPETITIVENESS IN NANOMANUFACTURING (CURRENT)
	What is known about how the United States compares to other nations in terms of investments in nanotechnology research and development, commercialization, and scale-up? What challenges to nanomanufacturing competitiveness does the United States face?
1:55 p.m.	**The Global Investment Profile in Nanotechnology— Comparing the U.S. to Selected Nations** Françoise D. Roure (10 min.) Mihail C. Roco (10 min.)
2:15 p.m.	Forum discussion (All)
3:10 p.m.	**BREAK**
3:20 p.m.	**Review of Profiles— Four Nanomanufacturing Areas**: Timothy M. Persons (5 min.) *Discussion of the four profiles: U.S. competitiveness and alternative views*
3:25 p.m.	The semiconductor industry – Discussants: Brian David Johnson, Michael Liehr, Celia Merzbacher (15 min.)
3:40 p.m.	Battery-powered vehicles – Discussants: Sarbajit Banerjee, Bill Canis (15 min.)
3:55 p.m.	Nano-based concrete – Discussants: Björn Birgisson, Hamlin M. Jennings (15 min.)
4:10 p.m.	Nanotherapeutics in medicine – Discussants: Joseph DeSimone, Scott E. McNeil (15 min.)
4:25 p.m.	Forum discussion (All)
5:30 p.m.	**RECEPTION**
	DAY TWO: WEDNESDAY, JULY 24, 2013
8:00 a.m.	ARRIVALS, COFFEE
8:30 a.m.	Welcome, Review of Day 1, and Preview of Day 2: Timothy M. Persons (5 min.)
	FUTURE ECONOMIC COMPETITIVENESS AND IMPORTANCE
	What is anticipated about how fast markets for advanced nanomanufacturing might expand? What are possible futures for U.S. competitiveness? In experts judgment, how important will U.S. competitiveness in nanomanufacturing be in the future?
8:35 a.m.	**Prospects for Nanotechnology and Economic Importance**: Mihail C. Roco (10 min.)
8:45 a.m.	**U.S. Nanomanufacturing: Issues and Practice, Looking Forward**: Manish Mehta (10 min.)
8:55: a.m.	**Issues in Innovation Policy**: Susan Offutt (10 min.)
9:05 a.m.	Discussants: Brian David Johnson, Matthew Nordan (15 min.)
9:20 a.m.	Forum discussion (All)
10:30 a.m.	**BREAK**

	ENHANCING COMPETITIVENESS IN THE YEARS AHEAD
	What are some steps that are being suggested, or that experts might suggest, to enhance the future of U.S. competitiveness in nanomanufacturing— including potential actions by the federal government as well as others?
10:40 a.m.	**Joining Manufacturing, Design, and Innovation**: Michael Molnar (10 min.)
10:50 a.m.	**Options for Enhancing Future U.S. Competitiveness**: Charles Wessner (10 min.)
11: 00 a.m.	**Strategies and Scenarios for the Years Ahead**: Sheila R. Ronis (10 min.)
11:10 a.m.	Discussants: John Ho, James M. Phillips (15 min.)
11: 25 a.m.	Forum discussion (All)
12:30 p.m.	WORKING LUNCH: NANOMANUFACTURING— REPORTS FROM PRACTITIONERS James M. Phillips, Joseph DeSimone, and Brian David Johnson
	OUTLOOK FOR THE ENVIRONMENTAL, HEALTH, AND SAFETY ASPECTS OF NANOMANUFACTURING
	What are concerns about nanotechnology products' EHS impacts? Do the concerns about nanotechnology impacts or approaches for managing those risks differ from those associated with other kinds of products?
1:30 p.m.	**Nanomanufacturing Workforce Health**: Paul Schulte (10 min.)
1:40 p.m.	**Nanomanufacturing and the Environment**: Lynn L. Bergeson (10 min.)
1:50 p.m.	Discussants: Tina Bahadori, Vicki L. Colvin, David Rejeski (20 min.)
2:10 p.m.	Forum discussion (All)
3:15 p.m.	**BREAK**
3:25 p.m.	Summary of Forum and Concluding Remarks (Timothy M. Persons and All)
4:25 p.m.	**END OF FORUM**

APPENDIX II: LIST OF FORUM PARTICIPANTS

Host	Gene L. Dodaro, Comptroller General of the United States.
Participants	George Allen, Former U.S. Senator and former Governor of Virginia.
	Tina Bahadori, National Program Director, Chemical Safety for Sustainability Research Program, Environmental Protection Agency, Washington, D.C.

	Sarbajit Banerjee, Associate Professor of Chemistry, College of Arts and Sciences, University at Buffalo, The State University of New York, Buffalo, New York.
	Lynn L. Bergeson, Managing Partner, Bergeson & Campbell PC, Washington, D.C.
	Bjorn Birgisson, Vice President for Research, KTH Royal Institute of Technology, Stockholm, Sweden.
	Bill Canis, Senior Analyst, Transportation and Industry Analysis Section, Resources, Sciences and Industry Division, Congressional Research Service, Washington, D.C.
	Vicki L. Colvin, Kenneth S. Pitzer-Schlumberger Professor of Chemistry; Professor of Chemical and Biomolecular Engineering, Rice University, Houston, Texas.
	Joseph DeSimone, Director, Frank Hawkins Kenan Institute of Private Enterprise, and Chancellor's Eminent Professor of Chemistry, University of North Carolina, Chapel Hill, North Carolina.
	Bart Gordon, Former Chairman, Committee on Science and Technology, House of Representatives, United States Congress; Partner at K&L Gates LLP, Washington, D.C.
	John Ho, Advanced Development Manager, QD Vision Inc., Lexington, Massachusetts.
	Hamlin M. Jennings, Principal Investigator, Concrete Sustainability Hub; Adjunct Professor, Department of Civil and Environmental Engineering, Massachusetts Institute of Technology, Cambridge, Massachusetts.
	Brian David Johnson, Futurist and Principal Engineer, Intel Corporation, Santa Clara, California.
	Michael Liehr, Executive Vice President of Innovation and Technology, Vice President for Research, College of Nanoscale Science and Engineering, State University of New York, Albany, New York.
	Scott E. McNeil, Director, Nanotechnology Characterization Laboratory for SAIC-Frederick and the Frederick National Laboratory for Cancer Research; Vice President, Science Applications International Corporation (SAIC), Frederick, Maryland.
	Manish Mehta, Director, Strategic Projects and Sustainability, Principal Investigator, NCMS-NSF Nanotechnology Commercialization Readiness Study, National Center for Manufacturing Sciences, Ann Arbor, Michigan.
	Celia Merzbacher, Vice President of Innovative Partnerships, Semiconductor Research Corporation, Research Triangle Park, North Carolina.

	Michael F. Molnar, Director, Advanced Manufacturing National Program Office,[72] and Chief Manufacturing Officer, National Institute of Standards and Technology, Gaithersburg, Maryland.
	Matthew Nordan, Vice President, Venrock, Cambridge, Massachusetts.
	Susan E. Offutt, Chief Economist, U.S. Government Accountability Office, Washington, D.C.
	Timothy M. Persons, Chief Scientist, U.S. Government Accountability Office, Washington, D.C.
	James M. Phillips, Chairman of the Board and Chief Executive Officer, NanoMech Corporation, Fayetteville, Arkansas.
	Robert Pohanka, Director, National Nanotechnology Coordination Office, Arlington, Virginia.
	David Rejeski, Director, Science and Technology Innovation Program, Woodrow Wilson International Center for Scholars, Washington, D.C.
	Mihail C. Roco, Founding Chair, Subcommittee on Nanoscale Science, Engineering and Technology, U.S. National Science and Technology Council; Senior Advisor, Nanotechnology, National Science Foundation, Arlington, Virginia.
	Sheila R. Ronis, Chair and Professor, Department of Management, Walsh College, Troy, Michigan.
	Françoise D. Roure, Chair, Committee on Technologies and Society, French High Council for Industry, Energy, and Technologies; Chair, Working Party on Nanotechnology, Organisation for Economic Cooperation and Development, Paris, France.
	Paul Schulte, Director, Education and Information Division, National Institute for Occupational Safety and Health, Centers for Disease Control and Prevention, Atlanta, Georgia.
	Charles Wessner, National Academies Scholar; Director, National Academy of Sciences Technology, Innovation, and Entrepreneurship Program, Washington, D.C.

APPENDIX III: LIST OF OTHER EXPERTS CONSULTED

This list includes experts interviewed in preparation for the forum as well as those who reviewed a final draft.[73]

Mostafa Analoui, Head of Healthcare and Life Sciences, Livingston Securities, New York, New York.

Robert D. Atkinson, President, Information Technology & Innovation Foundation, Washington, D.C.

Eric Barnes, Vice President for Finance and Operations, NanoMech Corporation, Fayetteville, Arkansas.

Jon Belkowitz, President, Intelligent Concrete, LLC, Freehold, New Jersey.

Arden L. Bement, Jr., Former Director of the National Science Foundation, Former Director of the National Institute of Standards and Technology; Director Emeritus, Global Policy Research Institute, Purdue University, West Lafayette, Indiana.

Roger Bonnecaze, Co-Director, NASCENT Center, The University of Texas at Austin.

John R. Bukowski, Asphalt Pavement Team Leader, Office of Infrastructure, Federal Highway Administration, U.S. Department of Transportation, Washington, D.C.

Thomas A. Campbell, Associate Director for Outreach; Research Associate Professor, Institute for Critical Technology and Applied Science, Virginia Polytechnic Institute and State University, Blacksburg, Virginia.

Gardner A. Carrick, Vice President, Strategic Initiatives, The Manufacturing Institute, Washington, D.C.

Shaun Clancy, Director and Regional Head, Product Regulatory Services, Evonik Industries, Essen, Germany.

Raymond David, Manager of Toxicology for Industrial Chemicals, BASF Corporation, Florham Park, New Jersey.

Neil Desai, Vice President, Strategic Platforms, Abraxis Bioscience/Celgene Corp., Los Angeles, California.

Stephen Ezell, Senior Analyst, Information Technology & Innovation Foundation, Washington, D.C.

Michael M. Fancher, Vice President for Business Development & Economic Outreach; Director, New York State Center for Advanced Technology in Nanoelectronics & Nanomaterials; Associate Professor of Nanoeconomics, College of Nanoscale Science and Engineering, State University of New York, Albany, New York.

Omid Farokhzad, Associate Professor, Department of Anesthesia, Harvard Medical School; and Director, Laboratory of Nanomedicine and Biomaterials, Anesthesia, Brigham and Women's Hospital, Boston, Massachusetts.

Georgene Geary, State Research Engineer, Organizational Performance Management, Office of Chief Engineer, Georgia Department of Transportation, Atlanta, Georgia.

Charles L. Geraci, Jr., Coordinator, Nanotechnology Research Center, National Institute for Occupational Safety and Health, Cincinnati, Ohio.

Steve Gordon, Staff Scientist, 3M Medical Department, 3M Company, Maplewood, Minnesota.

Piotr Grodzinski, Director, Office of Cancer Nanotechnology Research, National Cancer Institute, Bethesda, Maryland.

David Howell, Team Lead, Hybrid and Electric Systems, Vehicle Technologies Program, U.S. Department of Energy, Washington, D.C.

Dexter Johnson, Analyst, Cientifica Ltd., New York, New York.

Eric "Rick"Luebbe, Chief Executive Officer, EnerG2, Seattle, Washington.

Robert A. Lutz, Retired Vice Chairman, General Motors Company, Detroit, Michigan.

Thom Mason, Laboratory Director, Oak Ridge National Laboratory, Oak Ridge, Tennessee.

André Nel, Professor of Medicine, Pediatrics, and Public Health, University of California, Los Angeles, California.

Amy Prieto, Chief Executive Officer, Prieto Battery, Inc., and Associate Professor of Chemistry, College of Natural Sciences, Colorado State University, Fort Collins, Colorado.

Daniel Sarewitz, Co-Director, Consortium for Science, Policy & Outcomes; Associate Director, Center for Nanotechnology in Society, Arizona State University, Tempe, Arizona.

Nora Savage, Environmental Engineer, National Center for Environmental Research, U.S. Environmental Protection Agency, Washington, D.C.

Surendra P. Shah, Director, Center for Advanced Cement-Based Materials; Walter P. Murphy Professor of Civil Engineering, Robert R. McCormick School of Engineering and Applied Science, Northwestern University, Evanston, Illinois.

Phillip A. Singerman, Associate Director for Innovation and Industry Services, Office of the Director, National Institute of Standards and Technology, Gaithersburg, Maryland.

S. V. Sreenivasan, Co-Director, NASCENT Center, The University of Texas at Austin.

Ian Steff, Vice President of Global Policy and International Relations, Semiconductor Industry Association (SIA), Washington, D.C.

Gregory C. Tassey, Former Director and Senior Economist, Economic Analysis Office; former Associate Director for Innovation and Industry Services, National Institute of Standards and Technology, Gaithersburg, Maryland.

Suneel N. Vanikar, Concrete Pavement Team Leader, Office of Infrastructure, Federal Highway Administration, U.S. Department of Transportation, Washington, D.C.

Mark Verbrugge, Management Committee Member, U.S. Advanced Battery Consortium LLC. (USABC), Southfield, Michigan.

Cyrus Wadia, Assistant Director, Clean Energy and Materials R&D, White House Office of Science & Technology Policy, Washington, D.C.

Falan Yinug, Director, Industry Statistics & Economic Policy, Semiconductor Industry Association (SIA), Washington, D.C.

APPENDIX IV: FORWARD-LOOKING PROFILES OF FOUR NANOTECHNOLOGY INDUSTRIES

- **Nanotechnology and the Future of the Semiconductor Industry**
- **Nanotechnology and the Future of Battery-Powered Vehicles**
- **Nanotechnology and the Future of Nano-Based Concrete**
- **Nanotechnology and the Future of Nanotherapeutics in Medicine**

This appendix reproduces, verbatim, an excerpt from the "Reading Package" e-mailed to participants on July 14, 2013, in advance of the forum, which was held July 23-24, 2013, at GAO Headquarters, Washington, D.C.

References cited in the profiles are included in the List of References (app. VII).

Profile 1: Nanotechnology and the Future of the Semiconductor Industry

Overview

A semiconductor is the generic term for the various devices and integrated circuits that regulate and provide a path for electrical signals. As such, semiconductors are the foundation of the electronics industry, and today's semiconductors are manufactured using nanoscaled materials and processes. For example, production of a number of the components in semiconductors currently takes place at the nanoscale—that is, at scales of less than 100 nanometers (nm). In 2012, semiconductors with features spaced 22 nm apart and with layers just a few nanometers in thickness entered high-volume production.

The manufacture of semiconductors has several stages, with the main stages including the design of the semiconductor chip, the production of these chips on silicon wafers at fabrication plants (or “fabs”), and the testing and packaging of the semiconductors. As Lux Research defines the nanotechnology value chain (using three categories—nanomaterials, nano-intermediates, and nano-enabled products), semiconductors are intermediate products with nanoscale features that are components of finished goods (in this case, consumer electronic devices and other items) incorporating nanotechnology.[74] Industry experts we spoke with said that the diffusion of semiconductor chips with nanoscale features is pervasive and that this technology continues to evolve. They noted that the semiconductor industry continues to make devices at smaller and smaller sizes and this will continue to improve. However, they also explained that that this “scaling” (the downward trend in size) can only go so far with the current silicon-based materials. They anticipated that the current silicon-based technology might continue to evolve for perhaps another 7 years and there would then likely be a move to a new, as of yet undetermined, technology (new materials or new architecture, or both). Industry experts we interviewed said that while graphene, a nanomaterial, is being tested as a replacement for silicon in manufacturing nano-devices, other materials are also being tested, as are new chip architectures; the experts estimated that the successor to the silicon-based semiconductor chip could be 10 or more years away.

Current U.S. Investments and Competitiveness

Data provided by the Semiconductor Industry Association (SIA) show global semiconductor sales to have been about $292 billion in 2012. Figure 15 (below) shows the global market shares of semiconductor sales by the location of the headquarters of the firm selling the semiconductors. From this perspective, the United States had a 50 percent market share, meaning that firms whose headquarters were located in the United States accounted for half of all semiconductors sold worldwide in 2012. SIA also reported in 2012 that U.S. direct semiconductor jobs totaled an estimated 244,800. SIA noted that this figure includes both (1) the number of semiconductor jobs reported by the U.S. Bureau of Labor Statistics (BLS), and (2) an estimate by SIA of the number of jobs in the “fabless” semiconductor design sector—that is, jobs related to designing semiconductors but not actually manufacturing them.

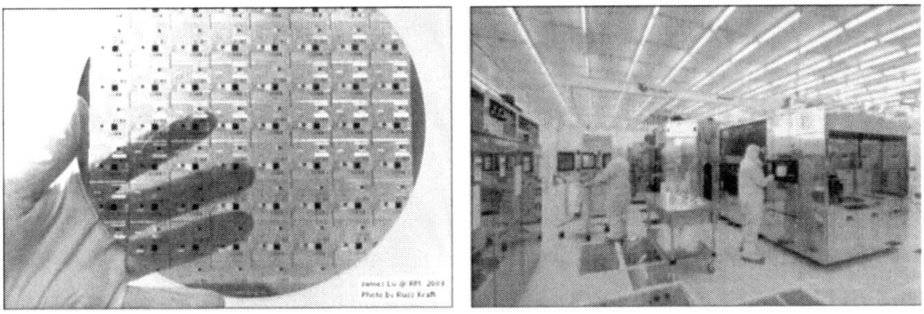

Sources: Rensselaer Polytechnic Institute, Troy, New York (left photo); College of Nanoscale Science & Engineering, State University of New York (right photo).

Note: The photograph on the left shows chips designed for 3D-interconnection, bonded to a glass (nonconductive) substrate, after removal of their silicon substrate. Each chip is designed to be stacked with other different chips in a way that allows a 3D vertical interconnection, thereby enabling 3D ICs (three-dimensional integrated circuits).

Figure 14. Glass Wafer with Multiple Semiconductor Chips, Each Having Nanoscale Features; View of a Semiconductor-Manufacturing Facility.

SIA also cited 2011 BLS data showing that U.S. semiconductor industry jobs stretch across almost every region of the country and into the majority of states. BLS reported that U.S. semiconductor jobs grew by 3.7 percent over the prior year, compared with lower growth (1.2 percent) for the broader economy. SIA called the semiconductor industry "the backbone of modern technology," suggesting that the industry may have a positive effect on job creation in other sectors of the U.S. economy.

Experts indicated that the degree of competitiveness of the U.S. semiconductor industry varied depending on whether the firm focused on design or other aspects of manufacturing. Industry experts we spoke with said that the United States is and will likely remain dominant in design—although they anticipated that U.S. dominance would be challenged by other countries. Other experts pointed out that while there are fabs operating in the United States, significant manufacturing capacity is located in Asia:

- One of these experts noted the decline in the manufacturing of semiconductor chips in the United States over the years. He said that today some semiconductor firms cannot afford the investments to stay up to date with the latest technology.

- Other experts said that it can cost at least $1 billion more over a 10-year period to operate a fab in the United States than in some overseas countries, primarily due to differences in tax and investment policies.

Because the tools needed for the semiconductor industry are so expensive—around $100 million in some cases—some private companies in the United States (as well as companies headquartered in other countries) are willing to partner with a publicly funded entity like the College of Nanoscale Science and Engineering (CNSE) of the State University of New York, a unique research, development, prototyping, and educational cluster for nanotechnology. CNSE is also the location of the headquarters of SEMATECH, a global consortium of major computer chip manufacturers, which, among other things, coordinates cutting-edge research and development projects.

Future Economic Importance

SIA's analysis of the global semiconductor industry shows a rise in worldwide sales from $292 billion in 2012 to $333 billion in predicted sales in 2016. In addition, MarketLine, an industry analysis firm, examines semiconductor sales from the perspective of where the semiconductors are purchased. When viewed this way, MarketLine has noted that while the U.S. semiconductor industry saw significant fluctuations in the value of semiconductors purchased in the United States from 2007 to 2011, the forecast for the next few years anticipates moderate growth (see Figure 16 below).

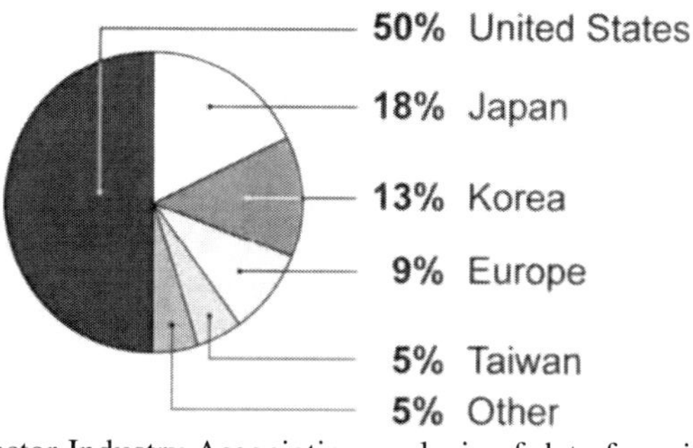

Source: Semiconductor Industry Association analysis of data from iSupply and World Semiconductor Trade Statistics.

Note: This chart is not directly comparable to similar charts in the other profiles.

Figure 15. Estimated Global Market Shares of Semiconductor Sales, 2012, by Headquarters Location of the Seller.

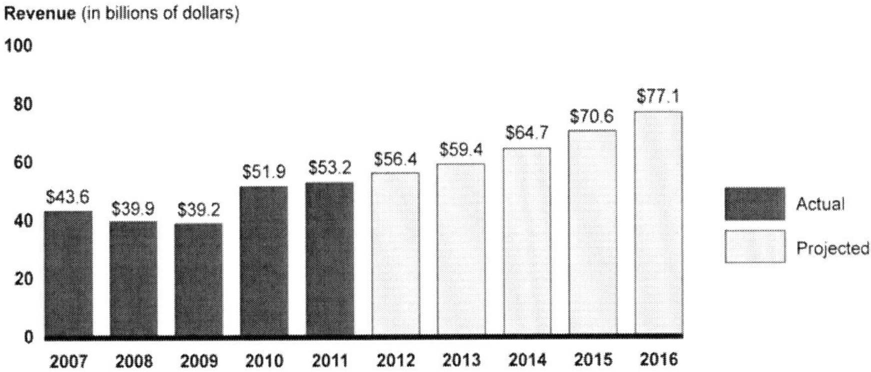

Source: MarketLine – Industry Report, October 19, 2012.

Figure 16. Actual and Projected Value of Semiconductors Purchased in the United States.

Enhancing U.S. Competitiveness in the Years Ahead

SIA and its member companies have identified what they believe are the "six central issues that will have a critical impact on the U.S. semiconductor industry's global competitiveness" in the coming years. These issues concern:

- corporate tax policy;
- export controls;
- environmental regulations;
- the semiconductor workforce, including science, technology, engineering, and mathematics (STEM) education and immigration policies;
- funding for basic research; and
- trade policy, including intellectual property protection.

Industry experts we interviewed about U.S. policies and challenges facing manufacturers mentioned many of these same issues. For example, these experts said they would like to see the federal research and development tax credit made permanent so firms could take a longer- term view. These experts also said federal funding for basic research is critical to develop new technologies and for the continuation of Moore's law, which posits that the number of transistors on a semiconducting chip doubles every 18 to 24 months. Another issue identified by other experts we interviewed concerns national strategy; specifically, these experts said that the United States does not have a cohesive strategy to assure leadership in the technology of the semiconductor industry. In comparison, in November 2012, a group of nano-electronic firms and institutes in Europe proposed a research and innovation program that aims to enhance large-scale semiconductor microchip manufacturing in Europe, increase nano-electronics-based revenues, and create a quarter of a million direct and indirect jobs.

In contrast with many of the experts we interviewed who proposed policies they believed would strengthen the semiconductor industry, the Organisation for Economic Co-operation and Development (OECD) recently proposed measures to support innovation but without an exclusive or targeted focus on manufacturing in general or the semiconductor industry in particular. Specifically, for the United States to maintain its "cutting edge," OECD (2012) recommended that it emphasize STEM education, preserve federal research and development funding, pursue labor policies to bring the long-term unemployed back into the labor market, enact some tax and fiscal policy changes, and so forth. Potentially, manufacturing would be supported by these policies, but so would other sectors of the U.S. economy.

Outlook for Environmental, Health, and Safety Aspects

Industry experts we interviewed pointed out that the semiconductor devices that are in large scale production today contain nanoscale features, but the finished chips themselves do not have freestanding nanoparticles, and therefore they do not have discrete engineered nanomaterials that can be released from the product. The experts said that generally, in producing semiconductor chips, manufacturers are trying to protect the semiconductor chips from the workers—hence the need for clean rooms. However, the experts also explained that the current semiconductor manufacturing process employs nano-sized particles in slurries that are not incorporated into the product, but are used to "polish" or "smooth" the surface of the semiconductor wafer during manufacturing. And some of the experts we interviewed acknowledged that there could be exposure to the compounds used to polish the silicon wafers. Other experts we interviewed agreed and noted that these polishing compounds are generally the only particles employees may come into contact with today in semiconductor manufacturing. While some experts believe these compounds can be handled appropriately, others stated that this issue could be studied further. Environmental, health, and safety (EHS) experts we spoke with pointed out that these slurries are used in high- volumes and that, because these slurries are used in clean room environments, the greatest concern for worker safety is where the materials are bulked, compounded, and piped into the clean rooms. Some experts also noted the potential for nanomaterial exposures from the use of single-wall carbon nanotubes and single-sheet graphene in semiconductors. While other experts pointed out that carbon nanotubes and graphene are not yet in large-scale production (and therefore have not been a workplace exposure issue), this illustrates that new issues may arise in the coming years. Finally, experts we interviewed stated that although the finished semiconductor chips themselves do not pose risks to the public, the effluents and wastes from semiconductor manufacturing processes must be carefully managed in order to prevent impact to the general public. And they stated that additional research is needed to be protective of human health and the environment.

Profile 2: Nanotechnology and the Future of Battery-Powered Vehicles

Overview

Researchers are using nanotechnology to improve the advanced batteries that power hybrids, plug-in hybrids, and fully electric vehicles (EVs).[75] These advanced batteries— often nanotechnology-based lithium-ion (Li-ion) batteries—work together with fuel or electricity, or both, as shown in figure 17. Nanoscale materials increase interactions between the anode and cathode, improving battery performance.

Battery-powered vehicles now represent about 3 to 4 percent of the U.S. and worldwide auto markets. Factors limiting demand for these vehicles include (1) the cost of an advanced battery, which increases the price of a battery-powered vehicle above that of a comparable all-gasoline car, and (2) the long battery-recharge times required by plug-in hybrids and EVs, and the EV's limited driving ranges. Potentially, nano-improved batteries will cost less than those currently available, have decreased recharge times, and provide the power to lengthen driving ranges.

Current U.S. Investments and Competitiveness

Experts we interviewed about nanotechnology and battery-powered vehicles— selected from government, industry, and academia—varied in their views about current levels of U.S. competitiveness and investments. One challenge to U.S. competitiveness, identified by some experts, consists of Asian dominance in Li-ion battery manufacturing. Current efforts to develop nano-improved batteries for vehicles center on Li-ion batteries—and the process for manufacturing Li-ion batteries for vehicles is similar to that used for other Li-ion batteries.

Although currently available figures for Li-ion battery manufacturing primarily reflect batteries for electronics, not vehicles—and although available estimates vary—as of 2010, they clearly indicate an overwhelming Asian dominance (Figure 18 provides one example).

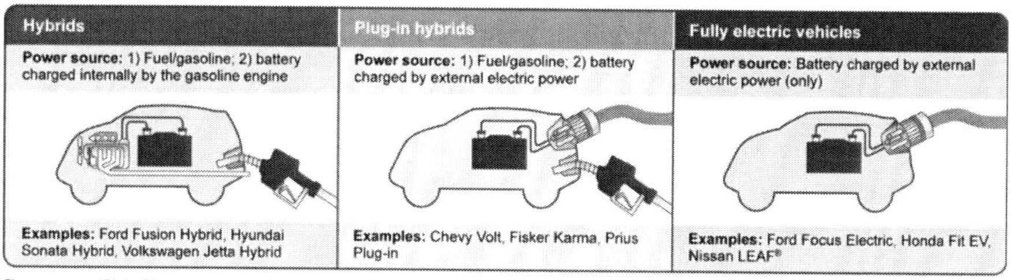

Source: GAO.

Figure 17. Three Main Types of Battery-Powered Vehicles.

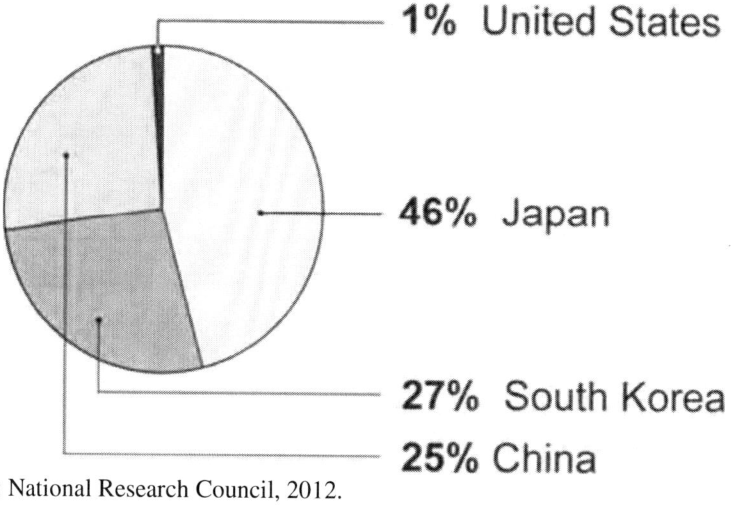

Source: National Research Council, 2012.

Note: This chart is not directly comparable to similar charts in the other profiles.

Figure 18. Estimated Global Market Shares, 2010, Based on Numbers of Lithium-ion Batteries Produced.

This dominance means an Asian advantage in terms of greater expertise in manufacturing processes for Li-ion batteries. A related challenge is that the number of competitors in this currently limited market makes it difficult for new U.S. (and other) companies to succeed against companies or nations with established expertise.

Two examples reflect these challenges:

- The California nano-battery startup Envia Systems locates its manufacturing in China and will "bring its technology to market" using partnerships and "selling manufacturing rights," rather than trying to locate a new manufacturing effort in the United States and compete with established Asian suppliers—especially given uncertain demand (LaMonica 2012).
- A123 was a U.S. start-up that located its manufacturing in Asia to gain expertise—then received federal support to establish a plant here. After going bankrupt in 2012, A123 was bought by a Chinese company, see, e.g., Bullis (2012). The plant (now named B456) is operated in Michigan.

While some experts emphasized that the nano-improved-batteries industry is in its infancy, viewed U.S. research as preeminent, and saw federal support as sizable, others were less positive. One said that other nations do more to "give their industries a leg up." Another said that in his view, U.S. investments are lower than—and in some cases, not as sustained as—other countries' and that some U.S. researchers now look to Asia for opportunities. And although private venture capital (or VC) is generally a key source of investment for U.S. startups, it may be insufficient in this case. One expert praised GM Ventures (which has funded startups such as Envia Systems, Inc., and Sakti3) because its staff identifies directions of interest to GM, to stimulate proposals. Yet another expert said VC generally looks for return on investment within 5 years, often resulting in a lack of long-term support for startups aiming for major innovations.

Future Economic Competitiveness and Importance

One expert anticipated two extreme future possibilities as equally plausible. The United States might

- become competitive or perhaps dominant because (1) the United States leads in research and (2) advanced manufacturing does not require large amounts of low-priced labor, or

- fail to become competitive (even with continuing research success), given the difficulties experienced thus far in attempting large-scale manufacture of Li-ion vehicle batteries.

Future U.S. manufacturing competitiveness will likely be important in this area, for three reasons, according to experts we interviewed: First, powerful new nano-improved batteries that are now in the research or idea stage, such as those illustrated in figure 19, are expected to enter the market.

Second, for targeted areas such as nano-improved batteries, the pace of R&D may quicken because researchers are increasingly able to design nanomaterials by controlling matter at the molecular level, designing batteries with specifically targeted characteristics.[76] Third—assuming these and perhaps other developments, such as rising gas prices and continued federal support for battery-powered vehicles—markets may expand, potentially along a trajectory such as that illustrated in figure 20.[77] Such a diffusion trajectory would, in turn, increase global demand for nano-improved batteries, but experts also cautioned that diffusion could be slowed if future gas prices remain steady or decline, thus reducing incentives for switching from gasoline-only models. (Two other potential obstacles concern possibly limited lithium supplies and possible grid overloads, although these are being or could be addressed.[78])

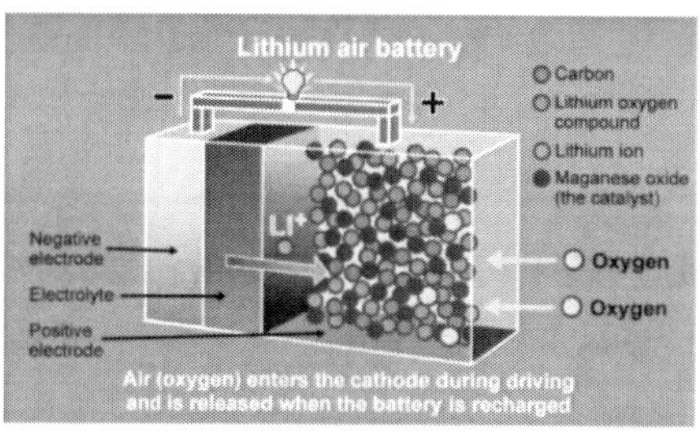

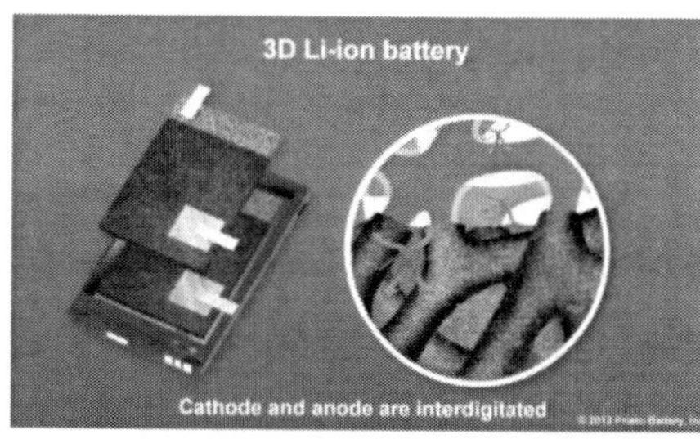

Source: Adapted from St. Andrews University, Scotland, UK illustration (left image); Prieto Battery, Colorado, USA (right image).

Note: On the 3D Li-ion battery illustration, the term *interdigitated* refers to an interlocked or interwoven design, in which copper nano-wires (anode) are coated with an ultra-thin electrolyte, then surrounded by a cathode slurry.

Figure 19. The Lithium Air and the 3D Li-ion Batteries, Illustrated.

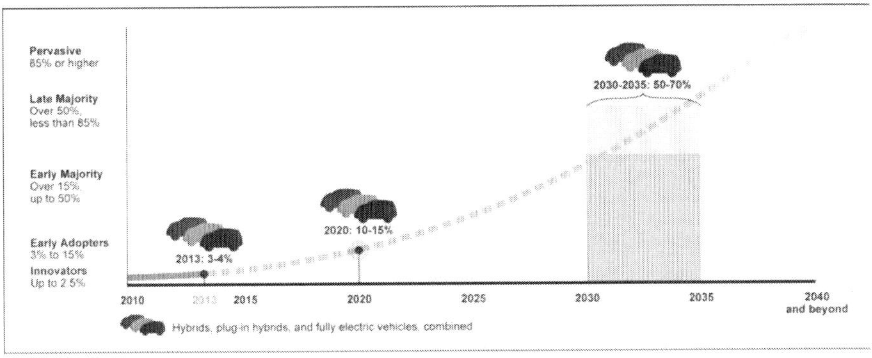

Source: GAO analysis based on (a) varied estimates and predictions from expert interviews and literature (Hirsch 2013; Price Waterhouse Coopers 2013), and (b) the cumulative diffusion of innovation curve suggested by Rogers (1962).

Note: There was some variation in experts' views. For example, one expert said that the 2020 estimate could range as high as 30% "if anticipated developments of advanced batteries come to fruition."

Figure 20. One View of Future Diffusion: Expanding Markets for Battery-Powered Vehicles.

Despite uncertainty about the future pace of diffusion, the prospect of a fast increasing market share prompted some experts to anticipate (1) an intensifying international race to develop and manufacture nano-improved batteries; (2) a focus on scaling up production of nano-engineered materials; and (3) an increased urgency about U.S. competitiveness issues in this area.

Finally, one interviewee foresaw a future opportunity for improved U.S. competitiveness—if new kinds of batteries, such as those illustrated in figure 19, require manufacturing approaches different from those now used for Li-ion batteries. (For example, the 3D battery would have a different manufacturing process than other Li-ion batteries because of its unique design.)

Enhancing U.S. Competitiveness in the Years Ahead

Generally, the experts we interviewed about battery-powered vehicles suggested that future U.S. competitiveness would be enhanced by:

- maintaining R&D tax credits;
- increasing federal support for research needed to take a new type of battery from lab scale to prototyping and on to small-scale pilot lines that show the ability to manufacture it;

- finding ways to "seed industry," such as bringing university and industry people together and increasing collaboration across the supply chain; and
- globally harmonizing regulations and standards for nano-improved batteries.

One interviewee said that it is also important for the United States to better understand "location decisions" and to anticipate "shake out" developments (in which some firms fail, leaving a smaller number of competitors) in order to stay on top of—or influence—trends.

Two specific examples of existing efforts that the experts we interviewed said might foster competitiveness, if expanded, are:

- *The new National Network for Manufacturing Innovation (NNMI), a federal program that collaborates with industries and state/local governments to fund new manufacturing-innovation institutes and "hubs."* One interviewee said the grass-roots aspect of NNMI would likely help avoid problems associated with prior funding of advanced battery manufacturing.
- *The proactive approach of GM Venture Capital, which outlines suggestions for future directions in which the company is interested.* This challenges researchers to design and propose technologies targeted to existing private-sector plans or initiatives. (For example, researchers studying carbon nano-products might benefit from knowing directions that nano-battery researchers are attempting to pursue or that existing VC is intending to fund.)

One expert emphasized that new efforts, such as these, should extend across the supply chain (from the manufacture of nanomaterials to finished consumer products) because competitiveness applies at every link and might be undermined by a single weak link. Potentially, a variety of other efforts might be discussed at the Forum.

Outlook for Environmental, Health, and Safety Aspects

Although the experts we interviewed specifically about battery-powered vehicles were either not very familiar with environmental, health, and safety (EHS) issues for nanotechnology manufacturing or not very concerned about risks, other experts we interviewed were selected for their knowledge about EHS issues. The EHS experts said that EHS risks could arise during manufacture and disposal of these batteries. To the extent that the nanomaterials remain contained in the battery during normal use, their impacts on consumers should be minimal, according to these experts. However, one expert stated that there are no systemic studies of the release of nanomaterials during either the battery use or disposal life-cycle phases. This expert stated that if the goal is to put many more electric vehicles on the road in the years ahead, a much more thorough analysis of life-cycle risks will be needed. Another expert stated that little effort is currently going into recycling or reclaiming material in the batteries.

Profile 3: Nanotechnology and the Future of Nano-Based Concrete

Overview

Concrete is the most heavily used construction material in the world—with about 5- billion cubic yards annually produced worldwide, according to the Portland Cement Association (2013)—and demand for it is expected to increase to meet the infrastructure needs of a growing global population. Nanomaterials can enhance the performance of the concrete used to construct this infrastructure. These materials might potentially result in roads, bridges, buildings, and structures that are more easily built, longer-lasting, and better-functioning than currently exist. For example, concretes using nanomaterials hold the promise of bridges and structures lasting for up to a century or more. Nanomaterials can enable this greater durability by strengthening the bonding ability of the cement paste that holds together a concrete mix.[79] They also can affect other physical and chemical properties of the concrete to improve its workability, allow it to cure more quickly, and resist cracking. Moreover, some nanotechnology-based concretes can beneficially convert polluting gases from combustion into substances less harmful to the environment. Figure 21 illustrates how nanotechnology-based concrete containing nano-titanium dioxide—a catalyst that uses sunlight to produce active forms of oxygen—reduces air pollutants by converting nitrogen oxides (NOx) to less harmful nitrate ions (NO_3^-). Such concrete was used in the construction of the I-35W bridge sculpture in Minneapolis, Minnesota.

Research is also being conducted on the use of nano-materials to produce concrete with self- sensing and -healing properties. These properties would, for example, allow highway engineers to monitor bridges and roads remotely and receive timely feedback on their structural conditions, thereby eliminating the need for time-consuming physical testing. Table 3 describes nanomaterials that can be used to enhance the performance of concrete and provides experts' views about current and future levels of market diffusion. This table also includes nanomaterials that can enhance the performance of asphalt, which is the predominant surface on U.S. roads supported with federal aid (GAO 2012b). However, this profile primarily focuses on nano-based concrete.

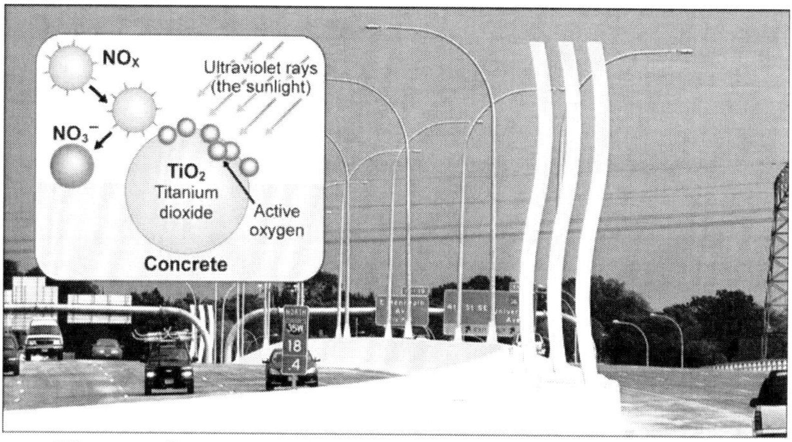

Sources: Minnesota Department of Transportation, David R. Gonzalez (photo); GAO analysis (inset).

Figure 21. Removal of Harmful Pollutants by Nanotechnology-Based Concrete in Sculpture.

Table 3. Examples of Nanomaterials in Concrete and Asphalt and Experts' Views on Market Diffusion in 2013 and 2020

		Views on diffusion of innovation	
Material	**Description**	**2013**	**2020**
Silica fume	Silica fume containing silicone dioxide increases the strength of concrete through improving the bonding characteristics of the cement paste that holds the mix together. In application, this material contributes to longer-lasting highways, bridges, buildings, and structures. A by-product of silicon metal production, silica fume has been used for several years. Silica fume particles are larger than 100	Market saturation[a]	Market saturation

		Views on diffusion of innovation	
Material	**Description**	**2013**	**2020**
	nanometers, the upper limit of the nanoscale; however, newer nano-silica materials sized less than this threshold are currently entering the market.		
Titanium dioxide	Titanium dioxide— through a photocatalytic process— enables exposed surfaces to absorb air pollutants and convert them to less harmful substances. This process allows for self-cleaning as rainwater washes away materials collected on the surface and results in concrete that is able to maintain its white color. This concrete may be used, for example, to construct roads and to create novel architectural features in buildings and structures. Titanium dioxide can also be used in asphalt pavement to absorb air pollutants through the hotocatalytic process described above. (See figure 21.)	Opinions varied from market entry[b] to market expansion	Opinions varied from market expansion[c] to market saturation
Nanoclays	Concrete containing nanoclays flows more easily and quickly gains strength. Concrete can thus be pumped to higher elevations to construct taller buildings than previously possible. Nanoclays provide another means to produce self-consolidating concrete which, in application, should reduce construction labor costs as fewer workers are needed to place the material.	Market entry	Market expansion
Nano-lithium	Nano-lithium works to seal and densify concrete. It is primarily used in the concrete-flooring industry to produce stronger and more impenetrable surfaces that should result in reduced maintenance and cleaning costs.	Market entry	Market expansion
Carbon nanotubes and nanofibers	Dispersed in concrete, carbon nano-tubes and nanofibers can improve strength and minimize shrinkage (or cracking). These materials can conduct electricity, which enables their use as a sensing mechanism. This self-sensing characteristic could allow, for example,	In research[d]	Market entry

Table 3. (Continued)

		Views on diffusion of innovation	
Material	**Description**	**2013**	**2020**
	remote monitoring for structural weaknesses or traffic volume.		
Bacteria	Bacteria can be used to make concrete that is self-healing. One concept currently being researched is bacterial mineral precipitation, which involves the mixing of dormant spores of bacteria in the concrete. If the concrete cracks and water seeps in, the bacteria consume the nutrients and produce calcite to fill the cracks.	In research	Market entry
Nanoparticles	Nano-alumina and nano-limestone are examples of nanoparticles. Such particles help accelerate hydration process— which is beneficial for quickly gaining strength and reducing the delayed strengthening of high volume fly ash concrete;[e] control shrinkage; reduce the heat produced during the curing process; and improve long-term strength.	In research	Market entry
Nanoparticles for asphalt	Nanoclays, nano-silica, nanomers, and carbon microfibers can be added to asphalt to improve pavement performance— such as making it more resistant to rutting and cracking— by affecting the chemical microstructure that influences physical behavior.	Market entry	Manufacturing expansion

Source: Selected experts and Birgisson et al., 2012.

[a] Market saturation generally means that the material is widely available and used.

[b] Market entry generally means that innovators and early adopters have developed and brought the material to the marketplace.

[c] Market expansion generally means that the material has expanded in the marketplace, beyond the innovators and early adopters.

[d] In research generally means exploratory and/or investigative activities that a business chooses to conduct with the intention of making a discovery that can either lead to the development of new products or procedures, or to improvement of existing products or procedures.

[e] Fly ash that results from the combustion of coal is the most commonly used mineral admixture in concrete. Mineral admixtures are often added to make concrete more economical, reduce permeability, increase strength, or influence other concrete properties.

Current U.S. Competitiveness

Experts offered differing views on U.S. global competitiveness as it relates to research and the commercialization and use of nanomaterials in concrete currently and in the future.

Research

The U.S.'s well-established research capacity enables its current competitiveness. Yet, according to some experts, other countries are likely to outpace the U.S. in research. As indicative of the strength of global competition, experts differed in their views on the technologies in which the U.S. will be competitively advantaged. For example, one expert anticipates that the U.S. will be competitive in research related to nano-TiO_2, -silica, and - admixture materials while another indicated the U.S. would be only slightly competitive in nano-TiO_2 research.[80] A third said the U.S. would not be competitive in nano-silica and - admixture research.

Commercialization and Use

With most nanomaterials in research or early market stages, some experts said that other countries are spending more resources than the U.S. to promote commercialization. One expert expects the U.S. to lead in two areas of commercialization: (1) nanoclays because this raw material is widely available in some states but limited in other parts of the world and (2) carbon nanotubes because of a strong domestic research and development base supporting the material.

Several factors may affect the commercialization and use of nanotechnology-based concrete materials. Participants in a 2007 workshop on nanotechnology for cement and concrete identified potentially key challenges.[81]

These challenges include the following:

- Owners may be unwilling to bear higher initial costs of using nano-based concrete as traditional concrete is a commodity material that is typically sold and placed under low bid contracts.
- Workers involved in making and placing concrete generally do not understand the material well; therefore, the mixture has to be easy to use and insensitive to mistakes and variability, and education must accompany changes in technology.
- Contracts tend to be inflexible with prescriptive specifications that often limit the acceptability of innovative approaches and materials

that although more costly in terms of up-front resources, may actually reduce total cost of ownership throughout the product's life cycle.

- Locally available raw materials, particularly aggregates, may not be suitable for use in concrete mixes and preclude use of nano-based concretes in some areas.

Future Economic Importance

The nano-materials shown earlier in table 3 will likely become more pervasive by 2020, highlighting the potential importance of nano-based concrete in an expanding global construction market. Chemical admixtures are one means to introduce nano-materials into concrete, making the chemical industry an important actor in the production of nano-based concretes. With a 15 percent share of chemical sales worldwide (of which chemical admixtures comprise a fraction), the U.S. is positioned to supply the chemicals necessary for some nano-based concretes—and benefit economically from their increased use. (See Figure 22.) Moreover, cement and raw material producers stand to similarly benefit from increased demand for concrete that may be driven by nanotechnology.

In addition to chemical and material producers, owners and end users of the constructed infrastructure could gain long-term economic benefits. For example, infrastructure constructed of nano-based concrete is expected to last longer and require less maintenance, leading to lower life-cycle costs and freeing resources that would otherwise be used for maintenance, repair, and reconstruction costs. In addition, economic benefits may be realized in terms of decreased user costs as motorists, for example, experience fewer delays on roads that are in better repair and less congested from maintenance and reconstruction work.

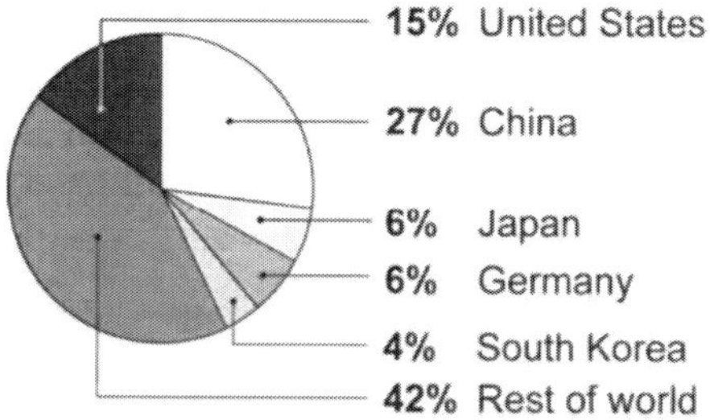

Source: Cefic–The European Chemical Industries Council, 2012.
Note: This chart is not directly comparable to similar charts in the other profiles.

Figure 22. Estimated Market Share of Chemical Sales by Country, 2011.

Enhancing U.S. Competiveness in the Years Ahead

Changing U.S. regulations, policy, and standards could increase demand for new nano-enhanced concrete, thereby stimulating innovation in the United States—and potentially strengthening the U.S.'s competitive position vis-à-vis other nations.

Regulations

Many experts we spoke with said that procurement regulations, such as those that state highway agencies use for construction projects, generally discourage use of proprietary products to ensure competition in the contracting process and require that awards be made to the lowest bidder. In addition, one expert noted that while these highway agencies might welcome new technologies, they are hesitant to use them absent sufficient evidence of past performance. As a result, contractors are reluctant to base their bids on use of proprietary materials—such as those needed to produce nano-concretes—because they may be more expensive and lack sufficient performance data. However, the Federal Highway Administration—which provides funding to states for highway construction and maintenance activities—recently changed its policy to allow states to use bid adjustment factors based on life- cycle-cost analyses in awarding federally funded pavement contracts, and experts said that some states are using procurement approaches that base contract award decisions on achieving lowest life-cycle cost. These actions indicate that states are beginning to take steps to address the challenges identified above by being more willing to bear higher initial costs of using nano-based concrete and writing contracts that provide greater flexibility for use of innovative approaches and materials.

Policy

Public policy decisions may affect the extent to which certain construction materials are used, as described by two examples. First, one expert said that policies to limit the carbon dioxide (CO_2) load on the environment could affect the pace at which new nano-based concrete materials are developed and implemented. In that cement production accounts for about 5 percent of global CO_2 emissions, using nano-based substitutes for traditional cement in concrete mixes can contribute to imposing a lower burden on the environment (Worrell et al., 2001). Second, as efforts evolve to allow greater use of private ventures to build, operate, and maintain infrastructure such as toll roads, concession winners might choose to use nanomaterials, higher initial cost notwithstanding, due to expectations of more economical performance over time.

Standards

Construction standards—typically incorporated into construction contract specifications—are important tools for defining acceptable characteristics of construction materials and the means by which they are used and tested. Having standards specific to nano-based concretes would help address the challenges identified above, for example, by providing information to workers handling the material and to designers in determining if locally available raw materials are suitable for use. Experts we spoke with said that while standards are necessary, changing them is a slow process because multiple stakeholders must reach agreement. Expediting change to standards, therefore, could help promote the use of innovative materials such as nano-based concretes.

Experts noted that compared to other countries, U.S. efforts to promote competitiveness are less apparent. For example, one expert said that China established a national technology center to improve its competitiveness and domestic production of high-value, nano-based construction products. Another expert said that while U.S. agencies such as the National Science Foundation promote strong fundamental research, the focus might not be on construction materials. This expert suggested a more prominent role for the National Science Foundation in construction materials research, which would include nanotechnology-based concrete materials.

Outlook for Environmental, Health, and Safety Aspects

Construction is labor intensive and generally done outdoors; therefore, as noted by the Transportation Research Board (Birgisson et al., 2010), nano-modified materials must be examined for their effects on health and the environment should they be released or leached from the concrete. Experts we spoke with who were familiar with environmental, health, and safety (EHS) issues expressed concerns, ranging from moderate to severe, about the effect of nanomaterials on workers' health, the public, and the environment and said that more information is needed on EHS implications. One expert explained that no guidelines are available for selecting clothing or other apparel to prevent dermal exposure to nanomaterials primarily because data available on the efficacy of existing protective clothing is minimal.[82] This expert added that the current hesitation in using nanotechnology could be attributed, in part, to experience using materials like asbestos that were useful but turned out to be very harmful to health. Other experts argued that industrial hygiene approaches work well for nanomaterials, but there is a need to increase awareness of nanomaterials among employers and workers. These experts were particularly concerned about secondary workers (those installing/applying nano-enabled products), who may not use protective equipment or strong industrial hygiene practices.

Profile 4: Nanotechnology and the Future of Nanotherapeutics in Medicine

Overview

According to experts, one of the most promising medical applications for nanotechnology is nanotherapeutics, the delivery of medicine using nanoparticles (particles having one or more dimensions on the order of 100 nanometers—100 billionth of a meter—or less). The potential of nanotherapeutics is the ability to target the delivery of drugs to specific cells—e.g., cancer cells—thereby reducing negative side effects. As one expert said, "nanotherapeutics has the potential to address problems in drug delivery for cancer and other diseases that cannot be solved using contemporary technologies." (See Figure 23.)

Although there are just a few nanotherapeutic drugs currently on the market, experts believe that more nanotherapeutic drugs are likely to come on the market in the next 7–10 years and that nanotherapeutic applications will continue to expand beyond cancer treatment to other conditions, such as infectious diseases, vascular disorders, and degenerative diseases. These statements are supported by a recent review that found that as of May 2011, there were about 150 nanotherapeutic drugs in various stages of clinical study with a handful already approved, and that while cancer treatment was the leading application for these nanotherapeutics, there were other applications in development (Etheridge et al., 2013). Generally, the drug development process for nanotherapeutic drugs is the same as for conventional drugs and includes drug discovery, preclinical testing, clinical trials, and regulatory review. (See Figure 24.) Overall, estimates for the process can average up to 15 years (GAO 2006).

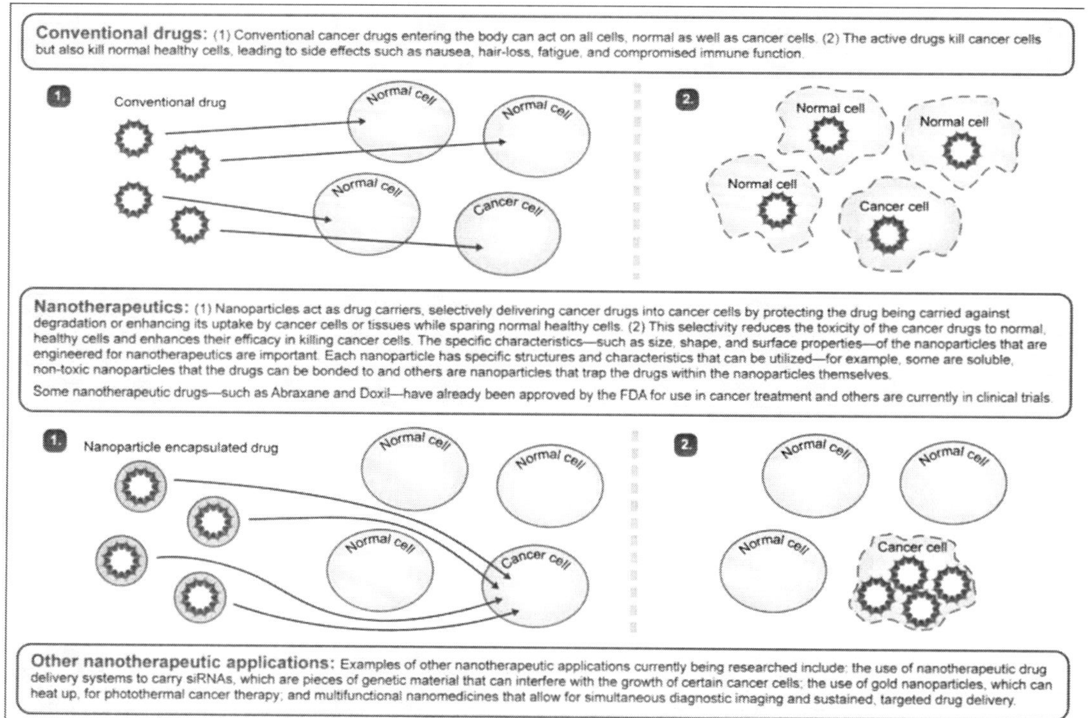

Source: GAO analysis based in part on Office of Cancer Nanotechnology Research 2010 and other literature.

Figure 23. Nanotherapeutics and Cancer Treatment.

Source: GAO based on industry estimates.

Figure 24. Overview of Drug Development Process.

Current U.S. Competitiveness

According to a European Commission, Joint Research Centre report, in 2008 the United States was the world leader in terms of patents for medical applications of nanotechnology (Wagner et al., 2008). (See Figure 25.) In terms of nanotherapeutic research, commercialization, and manufacturing, the experts we spoke with earlier this year said that the U.S. is dominant. They noted that while the volume of research from other countries has increased dramatically, U.S. research is among the best "if not the best in the world" for innovation and productivity, giving the U.S. a strong competitive advantage.

However, according to experts, the U.S.'s progress in nanotherapeutics could be hampered by challenges in the stages of the drug development process after drug discovery. Specifically, experts we interviewed said that while the U.S. is currently dominant in the commercialization and manufacturing of nanotherapeutics, these are potentially vulnerable areas because many of the efforts to commercialize nanotherapeutics are carried out by small companies. According to experts, small nanotherapeutic companies cannot typically sustain the costs of clinical trials and regulatory review. The experts said that in order to make it through these later stages of the drug development process, small companies need financial support from government, private investors, and/or larger companies; they said that these types of funding are limited. In particular, they noted that private investors can be reluctant to invest in new drugs—and not just nanotherapeutic drugs—because of the resources and uncertainty associated with gaining U.S. Food and Drug Administration (FDA) approval.[83]

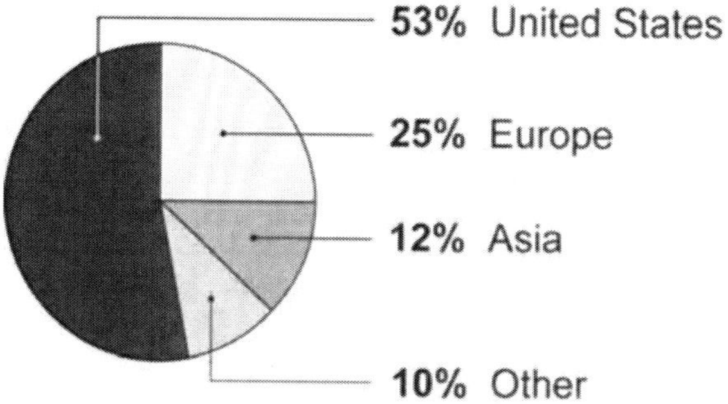

Source: GAO based on European Commission, Joint Research Centre data.
Note: This chart is not directly comparable to similar charts in the other profiles.

Figure 25. Percentage of Nanomedicine Patent Applications for United States, Europe, and Asia, 2008.

Table 4. Countries Making Significant Progress in Nanotherapeutics

North America	Europe[a]	Asia
United States	France	China
	Italy	Japan
	Germany	South Korea
	Russia	
	Switzerland	
	United Kingdom	

Source: Selected experts.

[a] Experts said that European countries sometimes act as one unit on nanotechnology and that other times countries act individually.

According to experts, there are about 10 countries, including the U.S., making significant progress in researching or manufacturing nanotherapeutics. (See table 4.) Experts also said that many of these countries are making large investments into research; infrastructure; and science, technology, engineering and math (STEM) education that will allow nanotherapeutics in those countries to make significant progress.

Future Economic Importance

According to a recent research report, the global nanomedicine sector reached $43.2 billion in 2010 and $50.1 billion in 2011, and is projected to grow to $96.9 billion by 2016 (BCC Research LLC 2012). According to this same report, the anticancer products segment of the global nanomedicine market is expected to reach $12.7 billion in 2016. Some of the experts we interviewed anticipated the following developments, which might tend to stimulate job growth:[84]

- New companies that facilitate early clinical trials, commercialization, and manufacturing.
- Expansion of contract research organizations skilled in nanotherapeutics.
- Continued growth of academic research (scientists and faculty) in nanotherapeutics.

Enhancing Competitiveness in the Years Ahead

According to experts, future U.S. competitiveness in nanotherapeutics will be affected primarily by the ability of the nanotherapeutic industry to secure sufficient funding, particularly for commercialization and manufacturing, and to have clear regulatory guidelines. For each of these areas, experts provided additional detail and examples of efforts that might enhance competitiveness.

The nanotherapeutic experts said that "funding drives everything" and that there are gaps in funding, particularly in the stages of the drug development process after drug discovery. To address these gaps, experts suggested that the federal government explore ways to fund nanotherapeutics at all stages, not just the research that happens during the drug discovery stage.

- *Federal funds for commercialization and manufacturing of nanotherapeutics.* Whether through expansion of SBIR grants or other mechanisms such as favorable tax incentives for small start-ups, experts suggested the federal government dedicate more funds to the commercialization and manufacturing of nanotherapeutics.
- *Creative funding mechanisms such as public private partnerships and funding from philanthropic organizations.* Experts said that other countries and regions—such as Europe and South Korea—have more diversified government funding mechanisms that allow for funding industry and industry-academia collaborations.
- *Private investments in nanotherapeutics.* According to experts, private investors are reluctant to devote funds to nanotherapeutics because of the time, money, and uncertainty associated with bringing drugs to market. Experts said that perhaps the federal government could create favorable incentives to encourage investors to take risks.

Experts said that clearly defined regulations and standards will benefit the development of nanotherapeutics. According to some experts, if U.S. regulations and standards are ambiguous, this could drive companies to other countries.[85] As with any new product or technology, it will be important to continue to balance the need to advance new nanotherapeutic technologies with the need to ensure the safety of new products. FDA said that in its regulation of nanomaterials, it "intends to ensure transparent and predictable regulatory pathways grounded in the best available science." Furthermore, FDA said it believes the current frameworks for evaluating the safety of FDA-regulated products are "sufficiently robust and flexible to be appropriate for a variety of products, including those that make use of nanomaterials."

- *FDA regulations.* FDA recently released guidance on the agency's approach for regulating nanotechnology products. FDA said it will consider releasing additional guidance as needed and that the agency is investing in its scientific capabilities regarding the regulation of nanotechnology.

- *FDA international coordination.* International coordination of regulatory efforts for nanotherapeutics is also important. According to some experts we interviewed, if other countries have less stringent regulations, they may have an advantage in bringing nanotherapeutics to the market faster. However, it should be noted that a 2011 analysis compared the FDA's review time for new oncology drugs to the European Medicines Agency's (EMA) review time and found that the FDA's median review time was shorter than EMA's (Roberts et al., 2011). Furthermore, as mentioned, it is paramount that the need to advance new technologies be balanced with the need to ensure the safety of new products. FDA said that it works with its international counterparts to "share perspectives and information on the regulation of nanotechnology products and their intended uses" and to develop nanotechnology-related technical standards.

Some experts also cited talent recruitment to STEM education as a challenge to the U.S.'s competitiveness; these experts said that unless the U.S. produces more graduates in STEM, we will begin to lag behind other countries.

Outlook for Environmental, Health, and Safety Aspects

Generally, the nanotherapeutic experts we talked with said they do not believe there is a reason to be significantly more concerned about nanotechnology products—such as nanotherapeutics—than other products associated with new technologies. Other experts in environmental, health, and safety (EHS) risks whom we spoke to stressed the importance of assessing risks across the life cycle of these products, including risks to those who administer the drugs and the fate of the materials. One expert noted that because nanotherapeutics will be more targeted than traditional drugs, the quantity needed will be reduced, a reduction that could potentially reduce exposure as well. The expert also said that it should be possible to develop a predictive model for nanotherapeutics based on the large amount of screening data obtained to date; FDA and other agencies have identified the need for this work in their strategic planning.

Appendix V: Scope and Methodology

This report summarizing discussions at the July 23-24, 2013, forum was produced through a multiphase process. The three main phases were: (1) selecting and inviting forum participants, who had a wide range of expertise and views, with the assistance of the National Academies; (2) developing a pre-forum Reading Package that included four nanomanufacturing industry profiles, and sending this to participants on July 12, in advance of the forum; (3) holding the forum, preparing an initial post-forum summary of the forum's discussions, submitting that summary to participants for their review, and considering and, as appropriate, incorporating the participants' responses and comments. Most of the initial summary was incorporated in the final draft of this report, which was sent to two outside experts for their review. The following sections discuss our approach to the three main phases outlined above, as well as covering other relevant issues, such as our selection and report of policy relevant topics, data reliability, the report's discussion of "considerations going forward," and disclosure.

Forum Participant Selection

To prepare for the Comptroller General forum, we contracted with the National Academies to assist us in participant selections. We met with National Research Council staff to help ensure balance and to assess potential conflicts of interest for forum participants.

In our initial discussions with the National Research Council staff, we agreed that forum participants should:

- as a group represent a range of backgrounds, experience, and knowledge in terms of representing (1) academia, business, government, and other sources, such as nonpartisan think tanks; (2) experience with research on and manufacturing of nanoparticles (including development of manufacturing methods); (3) diverse professional backgrounds with regard to knowledge in areas of science, economics, innovation policy, global competitiveness, manufacturing, foresight, and potential nanotechnology impacts on the environment, health, and safety;[86]
- from an individual perspective, be able to address topics such as (1) U.S. and other nations' investments in nanotechnology; (2) R&D

commercialization and scale-up; (3) the challenges to nanomanufacturing that the U.S. currently faces; (4) the future of nanotechnology and the economic importance of nanomanufacturing (e.g., whether or why future competitiveness in nanomanufacturing is important); and (5) what factors (e.g., STEM education or other infrastructure factors) are key to future U.S. competitiveness in nanomanufacturing; and

- further, include individuals so that the group would represent (1) diverse economic views on innovation policy, (2) knowledge about selected specific new initiatives that might enhance the future of nano- manufacturing competitiveness in the United States, and (3) diverse perspectives on the outlook for the environment, safety, and health implications of nanomanufacturing.

Our alternative criteria for defining experts in nanomanufacturing included the following (1) significant positions in organization(s) relevant to nanomanufacturing issues, including relevant manufacturing firms; (2) authorship of papers in professional journals or other substantial publications, relevant to nanomanufacturing issues; or (3) inclusion as speakers selected to appear on an expert panel or make key-note presentations relevant to nanomanufacturing issues. Each selected expert met one or more of these criteria.

To implement final selections for Comptroller General forum participants, first, the National Academies identified potential participants based on the criteria listed above. We and the National Academies then met again to discuss the National Academies' list of potential participants along with other participants whom we felt met the requisite qualifications. This phased strategy allowed the National Academies an opportunity to independently identify and internally discuss potential invitees before we shared suggestions for potential invitees. This phased strategy was employed to bring increased independence to the selection process; however, we made final determinations regarding participant selections.

Creation of Nanomanufacturing Industry Profiles, as Read-Ahead Material

To develop background for the forum participants, we selected four industry areas of nanomanufacturing for which we created expert-based, forward-looking profiles of nanotechnology and nanomanufacturing activity—with attention to four policy relevant topics. The nanomanufacturing areas we selected for expert-based profiling correspond to four of the nine non-defense industry areas listed by the National Nanomanufacturing Network (NNN) (see table 1 in the Introduction to this report).

In selecting these four nanomanufacturing areas for profiling, we consulted with a leading expert; our goal was to select four areas to (1) illustrate cross-industry variation in both the current pervasiveness of nanoproducts and the nanomanufacturing competitiveness of the United States in a global context and (2) to show that nanomanufacturing is relevant to industries that differ in terms of other factors, such as general levels of technological sophistication and rates of change. Based on application of these criteria and expert consultation, we chose (a) semiconductors, (b) nano-enhanced batteries for vehicles, (c) nano- based cement and concrete, and (d) drug delivery for cancer treatment.

For each of the four profiled nanomanufacturing areas, GAO analysts (1) conducted a preliminary literature review and created an initial short summary of what is known about nanotechnology and nanomanufacturing in the area, and (2) deployed a semi-structured interview protocol with four to eight expert-interviewees (per profile). The expert-interviewees who were selected to help us create the four profiles were selected from industry, government, and academia. The interview protocol included questions about the expert's views of current U.S. competitiveness, foresight questions about how nanotechnology research and manufacturing might develop in the coming years, opinion questions about ways of enhancing U.S. competitiveness, and questions on the outlook for environment, health and safety (EHS) aspects of nanomanufacturing. A final question asked each interviewee about any additional items, topics, or concerns they would like to share with us that had not been covered in our interview protocol.

After conducting the interviews, GAO analysts and specialists created a draft profile for each selected nanomanufacturing area, reflecting the experts' opinions as well as some material from literature. These draft profiles were further enhanced with added information on EHS aspects from a separate set of interviews with experts. Each of the four drafts was sent to two groups for review: (1) stakeholders within GAO representing different policy areas and (2) the original expert-interviewees. Comments from stakeholders and the expert-interviewees were incorporated in the profiles as appropriate. Finally, because experts interviewed about nanotherapeutics mentioned issues related to FDA, we sought responses to some questions from FDA, later provided

FDA with a draft of the nanotherapeutics profile for review, and considered FDA's comments.

The resulting four nanomanufacturing profiles were shared with invitees in advance of the Comptroller General's forum. The four profiles are also included in this report (app. IV presents the profiles). The profiles were not designed to be comprehensive or definitive; instead, they were developed to provide information that could stimulate discussion among an array of experts representing scientific, economic, policy, and business perspectives. The profiles were also intended to give readers information about the various ways in which nanotechnology and nanomanufacturing are being used and could potentially be used in future years as well as to help put forum findings in context.

A list of the experts whom we consulted (in addition to forum participants) regarding the four nanomanufacturing profiles and other issues related to the forum is included in appendix III. Additionally, we interviewed a recently retired government official with expert knowledge of the semiconductor industry, who asked not to be listed by name, and in obtaining background information (for example, suggestions for experts to interview), we and National Research Council staff members talked briefly with additional experts.

Comptroller General Forum and Participant Follow-up

The Comptroller General forum was held on July 23-24, 2013, at the GAO headquarters building. The forum was recorded and the discussion was transcribed. After the forum was held, we prepared a written summary of the discussions and sent this to participants asking for their comments. Briefly:

\

- Twenty-three of the 26 external (non-GAO) participants responded to our query.[87] Twenty-two of the 23 respondents concurred overall that the draft summary represented participants' remarks at the forum. The twenty-third respondent did not comment on the overall summary but did offer specific comments.
- Of the 22 concurring overall, one stated that worker safety issues merited further discussion in our initial summary. Other respondents had minor comments that suggested we expand the text, deepen certain areas, or include a specific point. We incorporated these comments as appropriate.

- Three forum participants did not respond with any comments on the summary.

Finally, we followed up with some forum participants to develop first-hand examples of some key points raised by participants in the meeting, references to relevant literature, and so forth.

Forum participants were not polled, and votes were not taken on positions at the forum. Rather, we reviewed meeting transcripts and interacted with participants after the meeting to summarize the discussion and prepare this report. In this report,

- The term "various participants" or "varied participants" generally means (1) we are aware that there was not evidence of a full agreement or consensus on the topics discussed, and (2) multiple participants made slightly different points that went to the larger point(s) we reported. In these cases after the main point is made, the report then usually details varied points made (including any countervailing/minority viewpoints if these were expressed at the forum).
- The term "multiple participants" generally indicates that more than one participant said something supporting a general point and that the views in support of the point were relatively uniform.
- Implicit forum consensus is generally termed "participants said" or "participants agreed" in view of the fact that 22 of the 26 external participants agreed with the initial written summary that we sent to participants after the forum was held.[88]

Selection and Report of Policy Relevant Topics

Based on a congressional request regarding nanotechnology manufacturing and initial discussions with experts, we identified relevant policy topics concerning nanotechnology manufacturing that we determined would be examined in a GAO Comptroller General forum. The forum agenda we created (which is included as app. I of this report), was designed to allow for considerable open discussion and flexibility.

Following the forum, we first identified four key topics for organizing material discussed, then sent our initial summary to participants, and subsequently promulgated this report. The four policy relevant topics we selected were (1) the future importance of nanomanufacturing; (2) U.S. investments and competitiveness in nanotechnology R&D and U.S. competitiveness in nanomanufacturing; (3) approaches to enhancing U.S. nanomanufacturing competitiveness in the years ahead; and (4) the outlook for nanomanufacturing and nanotechnology with respect to environmental, health, and safety (EHS) aspects.

Independence of Interviewees and Forum Participants

Because the Comptroller General forum was designed to address foresight issues involving nanotechnology manufacturing, it was necessary to interview experts and select forum participants who were on the cutting edge of nanotechnology. Given the need to obtain expertise from professionals in the field of nanotechnology and related science and policy areas, we could not avoid seeking the expertise of individuals whose professional reputations and economic interests might be affected by future developments in nanotechnology manufacturing.

Recognizing that the industry-area profiles we produced may reflect the views of interviewees who are in a position to benefit from increased government funding or other supportive efforts (as well as the views of government interviewees), we encouraged forum participants to maintain an awareness of the interviewees' perspectives when considering these profiles. For this reason, our pre-forum communications to forum participants included a disclaimer to this effect.

Additionally, the introduction to this report includes a similar disclaimer concerning forum participants. Notwithstanding this disclaimer, we took additional steps to exercise due diligence and to understand forum participants' potential conflicts of interest. We asked all forum participants to sign a form, which asked participants about their perspectives and circumstances. Specifically, we asked participants (1) whether they or their immediate family had any investments or assets that could be affected, in a direct and predictable way, by a decision or action based on the information or opinions they would provide to us; (2) whether they or their spouse receive any income or hold any organizational positions that could be affected, in a direct and predictable way, by the information or opinions they would provide us; (3) whether there were any other circumstances, not addressed in the two previous questions, that could be reasonably viewed by others as affecting participant viewpoints on the topics to be discussed. We received acceptable signed responses from all forum participants to these queries.

Data Reliability

Some of the quantitative data presented in this report are included, as background, in the Introduction or the four industry profiles. Data in these portions of the product are presented to provide context for readers. With respect to quantitative data in the main sections of the report, we did assess the reliability of budgetary figures and the data on country by country comparisons of public investments in nanotechnology and data on the authorship of scientific journal articles on the topic of nanotechnology. We found that the public investment data should be presented with a caveat and used graphical methods to convey uncertainty—because these data did meet our usual criteria for displaying data in a conventional manner.

With regard to public investment estimates and projections, different countries have different organizations and institutions that measure and administer public nanotechnology investments. These country-specific organizations and institutions do not use uniform measurement or technology categorization standards. Moreover, not all countries report such data publicly. Although the underlying figures for Germany, Japan, Russia, and China were adjusted for purchasing power parity to ensure that currency variations do not contribute to the non-uniformity in measurement, the data (as we explain) still lack precision. Consequently, measuring country by country public investments in nanotechnology is difficult.[89] Notwithstanding these issues we felt it was still important to convey the relative magnitude in public investment from some specific countries.

To visually communicate the lack of precision in international nanotechnology R&D funding projections, we used techniques consistent with those explained in *Best Practice Approaches for Characterizing, Communicating, and Incorporating Scientific Uncertainty in Climate Decisionmaking* (Morgan et al., 2009), in consultation with experts as described below.

Because of the uncertainty of the data underlying the projections, it would have been inappropriate for us to display specific end points in a bar chart representing these projections. Instead, we used shading techniques consistent with approaches used by Morgan and others as a way to graphically convey the uncertainty (that is, the lack of precision) in the underlying data. In each of the bars, the lack of specific end points is used to ensure that our report does not convey or imply a precise or specific amount of monetary investment for each country represented, because the non-uniformity of organizational structures and measurement techniques does not allow for such precision.

The types of imprecision were discussed with (1) one scientific organization and (2) two experts not affiliated with that organization. They stated that although figures for Germany and Japan were not precise, they were relatively more certain than data for public investments in Russia and China. This is why our graph shows projections for Germany and Japan in a deeper shade of blue. The presentation of data for Russia and China is complicated for the reasons listed above and, according to experts we spoke with, based upon a less transparent public disclosure of applicable information. For this reason, the data for Russia and China are relatively less certain than for Germany and Japan, and are presented in a lighter shade of blue to indicate less certainty.

Discussion of Considerations Going Forward

This non-audit engagement was designed to represent the viewpoints of experts who were selected to participate in the Comptroller General forum. We selected the experts with assistance from the National Academy of Sciences to ensure balance and representation of a wide range of significant viewpoints. We do not present formal recommendations addressed directly to executive branch agencies for their acceptance or rejection, nor do we present formal Matters for Congressional Consideration. However, we do discuss considerations going forward based on participants' views.

Disclosure

Forum attendees and other experts were informed that we would not directly identify individuals or their affiliations in association with specific comments (without their permission) and this product does not do so.

APPENDIX VI: EXAMPLES OF "TRANSFORMING GENERAL PURPOSE TECHNOLOGIES"

A 2005 assessment of economic transformations included the below table, which lists examples of transforming general purpose technologies and includes nanotechnology as the most recent example.

No.	Transforming general purpose technologies	Date[a]	Classification
1	Domestication of plants	9,000–8,000 BC	Process
2	Domestication of animals	8,500–7,500 BC[b]	Process
3	Smelting of ore	8,000–7,000 BC	Process
4	Wheel	4,000–3,000 BC[c]	Product
5	Writing	3,400–3,200 BC	Process
6	Bronze	2,800 BC	Product
7	Iron	1,200 BC	Product
8	Waterwheel	Early medieval period	Product
9	Three-masted sailing ship	15th century	Product
10	Printing	16th century	Process
11	Steam engine	Late 18th to early 19th century	Product
12	Factory system	Late 18th to early 19th century	Organizational
13	Railway	Mid 19th century	Product
14	Iron steamship	Mid 19th century	Product
15	Internal combustion engine	Late 19th century	Product
16	Electricity	Late 19th century	Product
17	Motor vehicle	20th century	Product
18	Airplane	20th century	Product
19	Mass-production, continuous-process factory[d]	20th century	Organizational
20	Computer	20th century	Product
21	Lean production	20th century	Organizational
22	Internet	20th century	Product
23	Biotechnology	20th century	Process
24	Nanotechnology[e]	21st century	Process

Source: Lipsey et al., 2005, 132.

Note: Lipsey et al., (2005, 98) define a general purpose technology as "a single generic technology, recognizable as such over its whole lifetime, that initially has much scope for improvement and eventually comes to be widely used, to have many uses, and to have many spillover effects."

The table notes below are quoted verbatim from this publication.

[a] "Many of these dates are approximate and based on rough estimates of when their use in the West became widespread enough for the technology to be identified as a general purpose technology from contemporary experience, although many were first innovated centuries and even millennia ago."

[b] "We [Lipsey et al.] include items 1 and 2 but not more modern agricultural developments because the domestication of plants and animals were truly generic developments with many uses that go far beyond food to such things as clothing, containers, shelter, transport, and power (many of which are still being worked out), while later innovations had a much narrower range of mainly agricultural uses."

[c] "There is little evidence regarding the origins of the wheel, but it was certainly not in use before the agricultural revolution and was in common use by about 3000 BC."

[d] "Although continuous process techniques began to evolve with the rationalization that followed the electrification of factories in the late 19th century, we date the emergence of mass production as a general purpose technology at Henry Ford's innovations in the first decade of the 20th century."

[e] "Nanotechnology has yet to make its presence felt as a general purpose technology, but its potential is so obvious and developing so quickly that we [Lipsey et al.] are willing to accept that it is on its way to being one of the most pervasive general purpose technologies of the 21st century."

APPENDIX VII: LIST OF REFERENCES

Bainbridge, William, Sims & Mihail C. Roco. (2006). "Progressive Convergence." Chapter 1 in *Managing Nano-Bio-Info-Cogno Innovations: Converging Technologies in Society*. Edited by Bainbridge, William Sims, and Mihail C. Roco. Netherlands, Springer, 1–8.

Berger, Suzanne. (2013a). "Production in the Innovation Economy: Preview of MIT's major new report on Manufacturing." Presentation at the National Academies: Washington, D.C., February 22.

Berger, Suzanne. (2013b). *Making in America: From Innovation to Market*. Cambridge, Massachusetts: The MIT Press.

BCC Research LLC. (2012). *Nanotechnology in Medical Applications: The Global Market*. Report Code. HLC069B. Wellesley, Massachusetts: BCC Research LLC, January.

Birgisson, Björn, Peter Taylor, Jamshid Armaghani & Surendra P. Shah. (2010). "American Road Map for Research for Nanotechnology- Based

Concrete Materials." *Transportation Research Record: Journal of the Transportation Research Board*, No. *2142*. Washington, D.C.: Transportation Research Board of the National Academies: 130–137.

Birgisson, Björn, Anal K. Mukhopadhyay, Georgene Geary, Mohammad Khan & Konstantin Sobolev. (2012). "Nanotechnology in Concrete Materials: A Synopsis." *Transportation Research Circular*. E-C170. Washington, D.C.: Transportation Research Board, December.

Boadway, Robin W. & David E. Wildasin. (1984). *Public Sector Economics*. Revised edition. Boston: Little, Brown, and Company.

Bradley, Jurron. (2010). "The Recession's Impact on Nanotechnology." *The Lux Research Analyst Blog*, February 4. (Based on the Lux Research report, *The Recession's Ripple Effect on Nanotech: State of the* Market Report. Boston, Massachusetts: Lux Research, Inc., June 9, 2009.)

Bullis, Kevin. (2012). "A123's Technology Just Wasn't Good Enough: Taking on Established Battery Makers and Making Electric Vehicles Mainstream Will Require Far More Radical Advances." *MIT Technology Review*, October 18.

Cefic aisbl. (2012). Facts and Figures 2012: *The European Chemicals Industry in Worldwide Perspective*. Brussels, Belgium: Cefic—The European Chemical Industries Council.

Christensen, Clayton M. (2012a). "A Capitalist's Dilemma, Whoever Wins on Tuesday." *The New York Times*, November 4.

Christensen, Clayton M. (2012b). "The New Church of Finance: Deeply Held Belief Systems and Complex Codes Must Be Changed." *Deseret News*, December 9.

Christensen, Clayton M. & Michael E. Raynor. (2003). *The Innovator's Solution: Creating and Sustaining Successful Growth*. Boston, Massachusetts: Harvard Business School Publishing Corporation.

Christensen, Clayton M., Richard Bohmer & John Kenagy. (2000). "Will Disruptive Innovations Cure Health Care?" *Harvard Business Review, 78* (5), September/October, 102–112.

Cientifica, Ltd. 2013. Data provided to GAO, March.

Clogg, Clifford C. (1979). *Measuring Underemployment: Demographic Indicators for the United States*. New York: Academic Press, Inc.

Council on Competitiveness. (2007). *Competitiveness Index: Where America Stands*. Washington, D.C.: Council on Competitiveness, January.

Etheridge, Michael L., Stephen A. Campbell, Arthur G. Erdman, Christy L. Haynes, Susan M. Wolf & Jeffrey McCullough. (2013). "The Big Picture on Nanomedicine: The State of Investigational and Approved

Nanomedicine Products." *Nanomedicine: Nanotechnology, Biology, and Medicine*, 9 (1), 1–14.

Executive Office of the President. (2012). *Report to the President on Capturing Domestic Competitive Advantage in Advanced Manufacturing*. Washington, D.C.: Executive Office of the President, July.

Freeman, Laura. (1996). "Job Creation and the Emerging Home Computer Market." *Monthly Labor Review*, August, 46–56.

GAO (U.S. Government Accountability Office). (2006). *New Drug Development: Science, Business, Regulatory, and Intellectual Property Issues Cited as Hampering Drug Development Efforts*. GAO-07-49. Washington, D.C., GAO, November 17.

GAO (U.S. Government Accountability Office). (2010a). *Best Practices: DOD Can Achieve Better Outcomes by Standardizing the Way Manufacturing Risks are Managed.* GAO-10-439. Washington, DC: GAO, April 22.

GAO (U.S. Government Accountability Office). (2010b). *Nanotechnology: Nanomaterials Are Widely Used in Commerce, but EPA Faces Challenges in Regulating Risk.* GAO-10-549. *Washington, D.C., GAO, May 25.*

GAO (U.S. Government Accountability Office). (2011a). *Small Business Innovation Research: SBA Should Work with Agencies to Improve the Data Available for Program Evaluation*. GAO-11-698. Washington, D.C., GAO, August 15.

GAO (U.S. Government Accountability Office). (2011b). *Technology Assessment: Neutron Detectors: Alternatives to Using Helium-3.* GAO-11-753. Washington, D.C., GAO, September 29.

GAO (U.S. Government Accountability Office). (2012a). *Nanotechnology: Improved Performance Information Needed for Environmental, Health, and Safety Research*. GAO-12-427. Washington, D.C., GAO, May 21.

GAO (U.S. Government Accountability Office). (2012b). *Materials and Practices for Improving Pavement Performance*. GAO-13-32R. Washington, D.C., GAO, November 14.

GAO (U.S. Government Accountability Office). (2013a). *Global Manufacturing: Foreign Government Programs Differ in Some Key Respects from Those in the United States*. GAO-13-365. Washington, D.C., GAO, July 25.

GAO (U.S. Government Accountability Office). (2013b). *International Regulatory Cooperation: Agency Efforts Could Benefit from Increased Collaboration and Interagency Guidance*. GAO-13-588. Washington, D.C., GAO, August 1.

Hassellöv, Martin, and Ralf Kaegi. (2009). "Analysis and Characterization of Manufactured Nanoparticles in Aquatic Environments." Chapter 6 in *Environmental and Human Health Impacts of Nanotechnology*, by Jamie R. Lead and Emma Smith. West Sussex, United Kingdom: Blackwell Publishing Ltd., 211–266.

Hirsch, Jerry. (2013). "Hybrid Vehicle Sales Speed Up: Fueled by Higher Gasoline Prices, a Growing Track Record for Reliability and a Wider Selection of Models, Hybrids' Share of the U.S. Auto Market Climbs to 4% and Could Double by Decade's End." *Los Angeles Times*, March 29.

Holman, Michael. (2007). "Nanotechnology's Impact on Consumer Products." (Slide presentation at a meeting organized by the Directorate- General for Health and Consumers, European Commission, October 25, 2007.) New York, New York, Lux Research, Inc.

IDTech Ex. (2013). "ORNL Research Paves Way for Larger, Safer Lithium Ion Batteries." *IDTechEx*, electricvehiclesresearch.com, March 27.

Khushf, George. (2007). "The Ethics of NBIC Convergence." *Journal of Medicine and Philosophy: A Forum for Bioethics and Philosophy of Medicine*, *32* (3), 185–196.

LaMonica, Martin. (2012). "Startup Envia Battery Promises to Slash EV Costs." *CNET Tech News*, February 26.

Langdon, David & Rebecca Lehrman. (2012). "The Benefits of Manufacturing Jobs." *ESA Issue Brief* #01-12. U.S. Department of Commerce, Economics and Statistics Administration, May.

Lipsey, Richard G., Kenneth Carlaw & Clifford Bekar. (2005). *Economic Transformations: General Purpose Technologies and Long-term Economic Growth*. New York, Oxford University Press, Inc.

Manykia, James, Jeff Sinclair, Richard Dobbs, Gernot Strube, Louis Rassey, Jan Mischke, Jaana Remes, Charles Roxburgh, Katy George, David O'Halloran & Sreenivas Ramaswamy. (2012). *Manufacturing the Future: The Next Era of Global Growth and Innovation*. McKinsey Global Institute, November.

MarketLine—Industry Report. (2012). October 19, 2012.

Mazov, Ilya, Vladimir Kuznetsov, Anatoly Romanenko & Valentin Suslyaev. (2012). "Properties of MWNT-Containing Polymer Composite Materials Depending on Their Structure." Chapter 3 in *Composites and Their Properties*. Edited by Ning Hu. Rijeka, Croatia: InTech, 37–60.

Mehta, Manish. (2010a). *2009 NCMS Study of Nanotechnology in the U.S. Manufacturing Industry. Final Report*. Prepared under NSF Award

Number DMI-0802026. Ann Arbor, Michigan, National Center for Manufacturing Sciences, August 23.

Mehta, Manish. (2010b). "*2009 NCMS Study of Nanotechnology in the U.S. Manufacturing Industry.*" Slides presented at NSERC Grantees Conference. Arlington, Virginia, December 7.

Mongillo, John. (2007). *Nanotechnology 101.* Westport, Connecticut, Greenwood Press.

Morgan, M. Granger, Hadi Dowlatabadi, Max Henrion, David Keith, Robert Lempert, Sandra McBride, Mitchell Small & Thomas Wilbanks, eds. (2009). *Best Practice Approaches for Characterizing, Communicating, and Incorporating Scientific Uncertainty in Climate Decisionmaking.* Synthesis and Assessment Product 5.2. Report by the U.S. Climate Change Science Program and the Subcommittee on Global Change Research. Washington, D.C., U.S. Climate Change Science Program, January.

Morse, Jeffrey D. (ed.). (2011). *Nanofabrication Processes for Roll-to-Roll Processing: Report from the NIST-NNN Workshop.* Workshop on Nanofabrication Technologies for Roll-to-Roll Processing. Seaport Convention Center, Seaport Boston Hotel, Boston, Massachusetts, September 27-28.

National Research Council. (2012). *Building the U.S. Battery Industry for Electric Drive Vehicles: Progress, Challenges, and Opportunities: Summary of a Symposium.* Washington, D.C., National Academies Press.

National Research Council. (2013). *Research Progress on Environmental, Health, and Safety Aspects of Engineered Materials.* Prepublication copy. Washington, D.C., National Academies Press, September 13.

NIOSH (National Institute for Occupational Safety and Health). (2013a). "*Occupational Exposure to Carbon Nanotubes and Nanofibers.*" Current Intelligence Bulletin 65. DHHS (NIOSH) Publication No. 2013-145. Cincinnati, Ohio, DHHS, April.

NIOSH (National Institute for Occupational Safety and Health). (2013b). *Current Strategies for Engineering Controls in Nanomaterial Production and Downstream Handling Processes.* DHHS (NIOSH) Publication No. 2014-102. Cincinnati, Ohio, DHHS, November.

ObservatoryNANO. (2012). "Public Funding of Nanotechnology." Seventh Framework Programme (2007 to 2013). European Commission.

Office of Cancer Nanotechnology Research. (2010). *Cancer Nanotechnology Plan.* National Cancer Institute, Center for Strategic Scientific Initiatives

(CSSI), National Institutes of Health. Publication No. 11-7704. Bethesda, Maryland: National Institutes of Health, November.

Organisation for Economic Co-operation and Development (OECD). (2007). "Competitive Regional Clusters: National Policy Approaches." OECD Policy Brief. Paris, France: OECD Publishing, May.

Organisation for Economic Co-operation and Development (OECD). (2012). *OECD Economic Surveys: United States 2012*. Paris, France, OECD Publishing, June.

Persons, Timothy M. (2013). "Comptroller General Forum on Nanomanufacturing: Overview." Slide presentation at the Comptroller General Forum on Nanomanufacturing. Washington, D.C., U.S. Government Accountability Office, July 23–24.

Pisano, Gary P. & Willy C. Shih. (2009). "Restoring American Competitiveness." Global Competitiveness in the World, Special Issue. *Harvard Business Review*, *87* (7-8) July-August, 114–125.

Price Waterhouse Coopers. (2013). *Charging Forward: PWC's 2012 Electric Vehicle Survey*.

Roberts, Samantha A., Jeff D. Allen & Ellen V. Sigal. (2011). "Despite Criticism of the FDA Review Process, New Cancer Drugs Reach Patients Sooner in the United States Than in Europe," *Health Affairs*, *30* (7), 1375– 1381.

Roco, Mihail C. (2013). "Global Investment Profile in Nanotechnology—Comparing U.S. to Selected Economies." Slide presentation at the Comptroller General Forum on Nanomanufacturing. U.S. Government Accountability Office, Washington, D.C., July 23.

Roco, Mihail C. & William S. Bainbridge. (2005). "Societal Implications of Nanoscience and Nanotechnology: Maximizing Human Benefit." *Journal of Nanoparticle Research*, *7* (1) February, 1–13.

Roco, Mihail C. & William S. Bainbridge. (2013). "The New World of Discovery, Invention, and Innovation: Convergence of Knowledge, Technology, and Society." *Journal of Nanoparticle Research*, *15* (9) September, 1-17.

Roco, Mihail C., Chad A. Mirkin & Mark C. Hersam. (2011). *Nanotechnology Research Directions for Societal Needs in 2020: Retrospective and Outlook*. New York, Springer Science+Business Media, 2011.

Rogers, Everett M. (1962). *Diffusion of Innovations*. Glencoe, Illinois: Free Press.

Roure, Françoise D. (2013). "The Global Investment Profile in Nanotechnology." Slide presentation at the Comptroller General Forum on

Nanomanufacturing. Washington, D.C., U.S. Government Accountability Office, July 23–24.

Rouse, Margaret. (2005). "Definition: Systems Thinking." SearchCIO.com, October; accessed September 12, 2013, http://searchcio.techtarget.com/definition/systems-thinking.

Shulaker, Max M., Gage Hills, Nishant Patil, Hai Wei, Hong-Yu Chen, Philip Wong, H.- S. & Subhasish Mitra. (2013). "Carbon Nanotube Computer." *Nature*, *501* (7468), September 26, 526–530.

Taylor, Peter, Krishna Rajan, Björn Birgisson & Tom Cackler. (2007). *Report of the Workshop on Nanotechnology for Cement and Concrete*. Center for Transportation Research and Education and the National Concrete Pavement Technology Center. Ames, Iowa: National Concrete Pavement Technology Center, September 5.

Tour, James. (2011). Testimony at hearing on "Oversight of the National Nanotechnology Initiative and Priorities for the Future" before the Subcommittee on Research and Science Education, Committee on Science, Space, and Technology, House of Representatives. Serial No. 112-15. 112th Congress, 1st Session, April 14, 2011. Washington, D.C., U.S. Government Printing Office, 2011, 54–58.

U.S. Government Accountability Office: please see GAO.

U.S. National Science and Technology Council. (2003). *Supplement to the President's 2004 Budget.*

U.S. National Science and Technology Council. (2005 through 2013). *Supplements to the President's 2006 though 2014 Budgets*. Arlington, Virginia, National Nanotechnology Coordination Office.

U.S. National Science and Technology Council. (2013). *The National Nanotechnology Initiative: Supplement to the President's Budget for Fiscal Year 2014*. Washington, D.C., Executive Office of the President, May.

Wagner, Volker, Bärbel Hüsing, Sibylle Gaisser & Anne-Katrin Bock. (2008). *Nanomedicine: Drivers for Development and Possible Impacts*. EUR 23494 EN. European Commission and Joint Research Centre. Luxembourg: Office for Official Publications of the European Communities.

Wessner, Charles W. (ed.). (2008). *An Assessment of the SBIR Program*. Washington, D.C., The National Academies Press.

Wessner, Charles W. (ed.). (2013a). *Best Practices in State and Regional Innovation Initiatives: Competing in the 21st Century*. Washington, D.C., The National Academies Press.

Wessner, Charles W. (Rapporteur). (2013b). *Building the Illinois Innovation Economy: Summary of a Symposium*. Washington, D.C., The National Academies Press.

Wessner, Charles W. (Rapporteur). (2013c). *Building the Ohio Innovation Economy: Summary of a Symposium*. Washington, D.C., The National Academies Press.

Wessner, Charles W. (Rapporteur). (2013d). *Strengthening American Manufacturing: The Role of the Manufacturing Extension Partnership: Summary of a Symposium*. Washington, D.C., The National Academies Press.

Wessner, Charles W. & Alan Wm. Wolff (eds.). (2012). *Rising to the Challenge: U.S. Innovation Policy for the Global Economy*. Washington, D.C., The National Academies Press.

Whitefoot, Kate S., and Steve Olson (Rapporteurs). (2012). *Making Value: Integrating Manufacturing, Design, and Innovation to Thrive in the Changing Global Economy: Summary of a Workshop*. Washington, D.C., The National Academies Press.

Worrell, Ernest, Lynn Price, Nathan Martin, Chris Hendriks & Leticia Ozawa Meida. (2001). "Carbon Dioxide Emissions from the Global Cement Industry." *Annual Review of Energy and the Environment*, Vol. *26*, 303– 329.

Yadav, Vinita, Jonathan D. Freedman, Mark Grinstaff & Ayusmen Sen. (2013). "Bone-Crack Detection, Targeting, and Repair Using Ion Gradients." *Angewandte Chemie International Edition*, *52* (42), October 11, 10977–11001.

Youtie, Jan, Maurizio Iacopetta & Stuart Graham. (2008). "Assessing the Nature of Nanotechnology: Can We Uncover an Emerging General Purpose Technology?" *Journal of Technology Transfer*, *33* (3), June, 315–329.

End Notes

[1] See Roco et al. (2011, xv).

[2] Mongillo (2007, 2).

[3] According to the U.S. National Science and Technology Council (2013, 3), the NNI is a federal initiative in which 27 units from various agencies and departments coordinate with each other and work toward achieving a future "in which the ability to understand and control matter at the nanoscale leads to a revolution in technology and industry that benefits society."

[4] Carbon nanotubes are nanoscale cylinders made of carbon and have extraordinary mechanical strength as well as favorable electrical properties.

[5] "Disruptive innovation" refers to a new technology that creates a new market (and a new value chain or "value network") and that ultimately, and often unexpectedly, overtakes an existing technology. See Christensen and Raynor (2003).

[6] Within the U.S. government, R&D refers to basic research activities and laboratory scale development; it does not include commercialization or commercial-scale development; see definition, accessed December 17, 2013, http://www.nsf.gov/statistics/randdef/fedgov.cfm.

[7] For purposes of this report, we define national competitiveness as the productivity with which a nation utilizes its set of institutions, policies, and human capital and natural endowments to produce goods and services, for the prosperity of its people. See also Council on Competitiveness (2007).

[8] The NNN describes itself as an alliance of partners from academia, government, and industry; these partners cooperate to advance nanomanufacturing in the United States.

[9] For purposes of this report, a megatrend is defined as a long-term, powerful, and evidence-supported process with a formative and transforming impact on the future. "Disruptive innovation" is defined as innovation that helps create a new market and value network, displacing an earlier technology—and may create jobs—as further discussed later in this report.

[10] Bainbridge and Roco (2006, 2) see the potential convergence of nanotechnology with biotechnology, information technology, and new technologies based in cognitive science (known as NBIC) as (1) constituting "a major phase change in the nature of science and technology" and (2) having "the greatest possible implications for the economy, society, and culture." Also on the topic of NBIC convergence, Khushf (2007) has written about upstream anticipation of ethical issues such as privacy and transparency. See also Roco and Bainbridge (2013).

[11] Lipsey et al., (2005, 98) provide an example: "the computer ... enabled the development of efficient, precisely controlled robots, which in turn enabled the restructuring of many factories along highly automated lines."

[12] Lipsey et al., (2005) list nanotechnology as a GPT, based on its potential (see app. VI of this report); similarly, Youtie et al., (2008) see nanotechnology as having potential to meet requirements for classification as a GPT.

[13] Roco and Bainbridge (2005; see especially table 1) describe a wide scope of nanotechnology's potential implications for society.

[14] Mentioned earlier, "disruptive innovation" is defined as innovation that helps create a new market and value network, displacing an earlier technology—and may create jobs. The disruptive process may occur over a few years or possibly decades; see, for example, Christensen and Raynor (2003) and Christensen (2012a). By contrast, these authors see "sustaining innovation" (which improves products for sale to the same consumers) as typically not creating jobs—and "efficiency innovation" (which essentially refers to new ways to produce the same products more efficiently for sale to the same consumers) as actually destroying jobs.

[15] As previously noted, within the U.S. government, R&D refers to basic research activities and laboratory scale development; it does not include commercialization or commercial- scale development.

[16] For purposes of this report, we define "offshoring" as a U.S. company's locating some or all of its manufacturing operations in one or more other countries for purposes of reducing costs.

[17] Other standards issues concern the environmental, safety, and health aspects of nanomanufacturing; those issues are discussed in a later section of this report.

[18] These two participants were selected, in part, for their expertise in international data concerning nanotechnology.

[19] One participant noted that private-sector R&D investments differ from public-sector investments in that private-sector investments tend to place greater emphasis on development (by contrast, public-sector investments place greater emphasis on research).

[20] That participant expressed the view that the three journals examined may tend to publish authors seen as being more prestigious or as having more prestigious affiliations, which might bias results in favor of the United States.

[21] The development of this method for nanomanufacturing medicines was based, in part, on earlier approaches used to produce semiconductors, and one participant suggested the potential importance of the experience of the semiconductor industry for nanomanufacturing in general.

[22] Specifically, the *Valley of Death* concerns the gap in support (typically, a lack of access to capital) that characterizes the transition of an invention from the point of validation in a laboratory environment to prototype demonstration in a non-laboratory environment (prior to acquisition by industry as a commercial product). An expert who commented on a draft of this report noted that another term associated with this gap is "the Badlands." He explained that "the Badlands" refers to "high barriers with restricted passageways through them." He further said that this gap in support is linked to what is essentially a disconnect between (1) the scientists and engineers who are skilled in conducting research and conceptualizing innovations and (2) venture capitalists, who, by contrast, are skilled in identifying and managing risks.

[23] Specifically, the *Missing Middle* typically concerns the stages of manufacturing development characterized by progression from proof of concept through production in the laboratory and the capacity to produce a prototype in a production-relevant environment.

[24] See Pisano and Shih (2009).

[25] Phase I SBIR supports the conduct of experimental or theoretical R&D, and as directed in the Small Business Reauthorization Act of 2000, applicants for Phase II SBIR awards are required to submit commercialization plans for technologies that are moving toward commercialization; see GAO (2011a).

[26] One participant, critical of the SBIR program, indicated that it may, in effect, take a small business out onto a bridge over the *Valley of Death* and then "push it off the bridge"—indicating that in this participant's view SBIR may help small businesses with the initial stages of the transition to commercialization but cuts off support before those businesses actually reach commercialization, leaving some without the follow-on support needed for further development.

[27] As one participant explained, this funding essentially "gave the United States a toehold" in the supply chain for nano-batteries for hybrids and electric vehicles (despite problems in establishing that industry in the United States).

[28] Participants said this development has been particularly problematic since the recent financial crisis and recession.

[29] In line with this participant observation, an expert (whom we interviewed prior to the forum) mentioned GM Ventures as a source of funding for commercializing nano-batteries for hybrids and electric vehicles.

[30] See also Wessner (2013a; 2013b; 2013c) and Wessner and Wolff (2012).

[31] As previously noted, for purposes of this report, we define "offshoring" as a U.S. company's locating some or all of its manufacturing operations in one or more other countries for purposes of reducing costs.

[32] A participant told us that this issue has been summed up as "the loss of companies that can make things will end up in the loss of research than can invent them," citing Berger (2013a).

[33] To illustrate this point, a participant said that the GlobalFoundries fabrication plant located in upstate New York employs about 1,000 engineers. GlobalFoundries has fabrication plants in the United States and elsewhere around the world.

[34] Semiconductor packaging involves encasing semiconductor electronic components in glass, metal, ceramic, or plastic—to protect against damage from, for example, corrosion or heat—and requires several important precision motions.

[35] Rockwell Automation's website describes the development and use of advanced robotics for this purpose; accessed October 30, 2013, http://www.rockwellautomation.com/anorad/solutions/semiconductor-packaging.page.

[36] U.S. companies, such as A123, that were formed to start manufacturing advanced vehicle batteries could not successfully compete with more experienced Asian firms— although low-skill labor costs are less important for advanced vehicle batteries than for consumer electronics. (One participant explained that batteries that power hybrid or electric vehicles may sell for between $8,000 and $18,000, and about 70% of this cost derives from the high-tech production of nano-engineered cells.)

[37] For example, this participant was hopeful about the prospects of Johnson Controls, which is headquartered in Milwaukee, Wisconsin.

[38] This view was questioned in a post-forum comment by a participant who pointed to a recent survey, which found that about three fourths of the manufacturing firms surveyed had filled all openings for core production workers within 3 months. Specifically, Berger (2013b, 185) reported that "most employers do not experience ... vacancies that last for three months or more," although 24 percent did report this. However, according to a spokesperson for this survey, it intends to represent manufacturers in general, and high- tech manufacturers were not broken out for separate analysis in the currently available report. The survey spokesperson noted that some current analyses *are* focusing on high- tech manufacturers, and a future publication may present information relevant to the issue above.

[39] One participant said that development of a unified system to describe nanomaterials could be the beginning of a "convergence and not divergence" of efforts regarding a regulatory framework.

[40] ISO is the International Organization for Standardization.

[41] Early *manufacturing* readiness levels range from the identification of basic manufacturing implications through developing a manufacturing proof of concept. For specific definitions of manufacturing readiness levels, see GAO (2010a), especially appendix II of that report.

[42] A fifth NSF-funded center related to nanomanufacturing is the Center for Nanomanufacturing Systems for Mobile Computing and Mobile Energy Technologies (NASCENT). This center has a different focus from the other four and is described in a later section of this report.

[43] One participant described this situation as approaching "technological war."

[44] The term "standard" can have different meanings. For example, it may be used to define a technical requirement or refer to a regulatory requirement (see GAO 2013b, 1; footnote 2). We use "basic standards" here to refer to technical characterization of materials and products and naming and measurement conventions, as opposed to regulatory standards related, for example, to environmental, safety, and health considerations.

[45] See GAO (2013b).

[46] As one participant said, education is mainly a responsibility of state and local governments, but it is also supported by federal policies and programs.

[47] See Clogg (1979, 223).

[48] In a similar vein, one participant expressed the view that R&D itself should not be overly targeted in certain directions by the government. Doing so could result in overlooking the best research directions.

[49] "Friction" is defined here as resistance encountered when pursuing a commercialization objective. To illustrate this concept using the example of innovative Internet services, there is less friction in starting up a firm offering new Internet services today than there was previously due to the advent of cloud computing, which has commoditized IT resources.

[50] See Morse (2011) for a discussion of roll-to-roll manufacturing.

[51] To date, CNSE has invested over $14 billion in facilities, including clean room space with state-of-the-art tools and equipment. A single piece of equipment in a clean room production facility could cost over $100 million.

[52] Although CNSE is not a federal program, it (like many other universities) receives federal funding for various specific projects or activities, such as research.

[53] According to the manufacturing.gov website (accessed November 19, 2013), "many technologies rooted in U.S. research fail to mature to full scale-up and commercialization in domestic factories. As documented by the National Science and Technology Council, 'A gap exists between R&D activities and the deployment of technological innovations in domestic production of goods,' contributing significantly, for example, to the disturbing and still-growing trade deficit in advanced technology products."

[54] The first NNMI institute, at a pilot level, was established in 2012, and three additional institutes (all of which are said to involve nanomanufacturing in some way) are planned for 2014.

[55] Private-sector partners might include large manufacturing companies, small and medium-sized enterprises, and start-ups.

[56] Regional manufacturing ecosystems, sometimes referred to as "clusters," are seen as enabling interaction among firms engaging in similar or complementary activities, research universities and institutes, vocational training schools, a well-trained labor force, financiers, parts and materials suppliers, marketing channels, and legal specialists. They may offer a well-developed transportation and communication infrastructure. Potentially, these varied elements help make the whole greater than the sum of its individual parts, through stimulating innovation and the development of small and medium-sized enterprises (SMEs).

[57] Many products not typically conceptualized as high tech are being manufactured in new ways—for example, textiles and athletic gear, such as running shoes. Additionally, as presented, the MDI nexus is actually broader than "making things" because of the connection between manufacturing and services; for example, innovative services may be associated with new technological developments or new products. See Whitefoot and Olson (2012).

[58] Systems thinking, which was developed in the 1950s at the Sloan School of Management at the Massachusetts Institute of Technology, has been applied by others; for example, a forum participant cited Edward W. Deming's use of this approach. See also Rouse (2005).

[59] See Christensen (2012a).

[60] For a review of how certain other countries are approaching innovative manufacturing, see GAO (2013a).

[61] Two examples are the X Prize and the Defense Advanced Research Projects Agency (DARPA) Grand Challenges.

[62] Another relevant program might be the Manufacturing Extension Partnership (MEP), which is currently "seeking to evolve beyond its traditional support for lean manufacturing to increase the innovation capacity of the nation's small and medium manufacturers" (Wessner 2013d, xiii-xiv).

[63] For a brief description of TTIP, accessed January 8, 2014, see http://www.whitehouse.gov/the-press-office/2013/06/17/fact-sheet-transatlantic-trade-and- investment-partnership-t-tip.

[64] See Langdon and Lehrman (2012, 1-2).

[65] In November 2013, NIOSH published guidance on engineering controls for nanomaterial production (see NIOSH 2013b).

[66] See GAO (2012a).

[67] The precautionary principle recognizes that government intervention beyond that normally justified by scientific evidence may be warranted if there are signals that a possible threat may, if unchecked, seriously harm the population.

[68] The manufacturing process known as 3-D printing has been defined as "build[ing] layers to create a three dimensional solid object from a digital model," accessed January 13, 2014,

http://whatis.techtarget.com/definition/3-D-printing-rapid-prototyping-stereolighography-or-architectural-modeling.

[69] The OECD Working Party on Nanotechnology (WPN) advises governments on emerging policy issues related to the responsible development of nanotechnology and promotes international cooperation to facilitate research, development, and responsible commercialization of nanotechnology.

[70] Early *technology* readiness levels concern the transition from scientific research to applied research and proof-of-concept validation; see GAO (2011b, 36).

[71] As explained in a previous section, early *manufacturing* readiness levels range from the identification of basic manufacturing implications through developing a manufacturing proof of concept. For specific definitions of manufacturing readiness levels, see GAO (2010a, app. II).

[72] The Advanced Manufacturing National Program Office is an interagency effort that administers the National Network for Manufacturing Innovation (NNMI).

[73] For additional details, please see appendix V.

[74] See Holman (2007).

[75] This profile focuses on the three main types of battery-powered vehicles shown in figure 6. We do not include "micro-hybrid" vehicles that reduce or eliminate use of the gasoline engine while idling and, in some cases, while running downhill.

[76] This is in contrast to discovering nano-substances and then identifying useful ways to apply them. For example, Oak Ridge National Laboratory intentionally designed a solid lithium nanomaterial that "could ultimately yield batteries . . . 5 to 10 times more powerful than current versions." (*IDTech Ex* 2013).

[77] U.S. government support has included (1) funding for R&D and in some cases, commercialization; (2) emissions requirements for vehicles sold here; and (3) in selected instances, federal tax credits to buyers of hybrid, plug-in hybrid, and electric vehicles. Logically, some of this support would aid many nations' efforts, not just U.S. efforts. Earlier this year, the President urged Congress to consider targeting part of the funding previously appropriated for energy initiatives ($2 billion) to promote advanced vehicles and lower their cost over 10 years by supporting research.

[78] The potential for limited lithium supplies, in the future, is being addressed in part by pursuing mining efforts, possible synthetic alternatives, and recycling plans. Ideas to avoid future grid overload include demand-related pricing and electronic feedback to drivers.

[79] Concrete consists of a cement paste made of water and cement that binds a mix of sand and aggregates. Cement paste hardens through a chemical reaction called hydration and gains strength to form the rock-like mass known as concrete.

[80] An admixture is a material—other than water, aggregates, cementitious materials, and fiber reinforcement—used as an ingredient of a cementitious mixture to modify its freshly mixed, setting, or hardened properties and that is added to the concrete before or during its mixing.

[81] This workshop was sponsored by the National Concrete Pavement Technology Center and the National Science Foundation, in cooperation with the Nanoscale Science, Engineering, and Technology Subcommittee of the U.S. National Science and Technology Council, through the National Nanotechnology Coordination Office (Taylor et al., 2007).

[82] The main reason for concern is that nanomaterials are very fine in nature and hence they have very large surface areas. In general, the more surface area increases, the higher the chances of reaction and exposure.

[83] The FDA is responsible for overseeing the safety and efficacy of drugs and biological products sold in the United States. FDA's review process involves evaluating scientific and clinical data to determine whether a drug meets statutory and regulatory standards for safety and effectiveness, manufacturing and controls, and labeling. For example, sponsors must provide "substantial evidence" of effectiveness for the claimed indications of the drug in order for FDA to approve the drug. FDA encourages sponsors to establish early interactions with FDA.

[84] According to the U.S. Bureau of Labor Statistics, in 2012, the overall U.S. pharmaceutical and medicine- manufacturing industry was estimated to employ almost 270,000 people.

[85] However, one expert said that because of the complexity of manufacturing nanotherapeutics, the offshoring of nanotherapeutics will be slower than for other types of drugs.

[86] We also decided that the group as a whole should have representation across our four nanomanufacturing profile areas: (a) semiconductors, (b) nano-batteries for vehicles, (c) nano-based cement/concrete, and (d) drug delivery for cancer treatment, as well as other areas.

[87] In addition to the 26 external participants, GAO's Chief Scientist and Chief Economist participated as experts (see app. II).

[88] As stated above, of the four external forum participants not counted as agreeing with the initial written summary, one responded by providing specific comments but did not specifically indicate agreement or disagreement overall; the other three did not respond to our inquiry.

[89] For this reason, this report includes a participant's suggestion for increased international cooperation on standards and data collection and reporting in order to improve our knowledge of the economics of nanotechnology.

INDEX

C

D

E

H

I

J

K

L

M

N

O

P

Q

R

S

T

U

V

W

Y